ISO a practical guide

for SMEs

AF880919

ISO 14001:2015

Environmental management systems

ISBN 978-92-67-10679-3

ISO copyright office
CP 401 • CH-1214 Vernier, Geneva
Tel. +41 22 749 01 11
Fax. +41 22 749 09 47
E-mail copyright@iso.org
Web www.iso.org

About the Author

Susan L.K. Briggs is the Head of Delegation of the USA national standards body to ISO/TC 207 on Environmental Management, and member of its subcommittee on Environmental Management Systems, serving on the working group that developed the 2004 versions of ISO 14001 and ISO 14004. She participated on the Joint Technical Coordination Group task force that developed ISO Annex SL, the core management system requirements and related guidance now used in all ISO management systems standards. She was elected by her TC 207/SC 1 colleagues to the leadership role of the international working group (WG 5) responsible for the 2015 revision of ISO 14001.

About the Reviewers

Johan G. Nel, Professor of Environmental Management, is the founder and Director of the Centre for Environmental Management at the North-West University, South Africa. He is a founding member and current chair of the South African TC 207 mirror committee and represented South Africa in the drafting of ISO 14001:2015.

Horacio Martirena is the director of a consultancy company specializing in management systems. Since 1991, he has been a member of the TC 176 working groups developing ISO 9001, and since 1999 a member of the TC 207 groups developing ISO 14001. He represented TC 207 in the working group that developed Annex SL and acted as vice-convenor of the working group responsible for the development of ISO 14001:2015.

Stefan Larsson is a consultant, board member and mentor of a consultancy working with organizations around the world on management systems and other ways of making organizations more sustainable. He was the chief technical expert for Sweden in TC 207 on the development of ISO 14001:2015 and is now working in the same capacity in PC 283, developing ISO 45001.

Acknowledgements

ISO published this handbook to support the implementation by SMEs of ISO 14001, *Environmental management systems — Requirements with guidance for use*, which was developed by ISO/TC 207.

ISO wishes to thank the author Susan L.K. Briggs, in addition to Johan G. Nel, Horacio Martinera and Stefan Larsson for providing feedback on the handbook.

Views expressed in this publication are those of the authors and do not necessarily reflect those of ISO.

Contents

Page

Foreword

Awareness of environmental problems is growing in most countries and it is believed that current development patterns cannot be sustained in the long term. Governments are increasingly enacting legislation aimed at protecting the environment, and customers are requiring their suppliers to incorporate best practices and demonstrate compliance with environmental requirements.

The publication of the ISO 14001 standard for environmental management systems (EMS) has proved to be very successful, as it is now implemented in more than 159 countries and has provided organizations with a powerful management tool to improve their environmental performance. More than 324 000 organizations have been certified worldwide against ISO 14001 at the end of 2014, which is an increase of 7 % compared to 2013. Many companies have improved their operations by reducing the adverse impact their activities, processes, products and services have on the environment using a systematic approach embodied in ISO 14001.

The benefits of positively addressing environmental issues therefore not only cover the protection of the environment, but are also linked to business performance and profitability. These can include improving the corporate image, enhancing access to export markets, and improving relationships with customers, regulators, the public and other stakeholders, etc.

One major advantage of implementing ISO 14001 is that it can be done in an organization of any size or type, since the requirements of an EMS are the same for all – although the manner of implementing one will vary according to the size and activity of the organization.

Experience shows that small and medium-sized enterprises (SMEs) can implement an effective EMS and realize a variety of benefits. However, EMS implementation can present some challenges. This guide aims to help organizations understand the requirements for an environment management system and for them to identify areas for improvement. It will be of value in preparing the organization for third party certification of its EMS should the organization wish to do so.

We hope that this revised guide to help achieve the benefits of ISO 14001:2015 will be of practical use to small businesses whatever their activity and wherever they may be, but especially in developing countries and economies in transition.

Nicolas Fleury
Acting Secretary-General
ISO

Introduction

This guide guides the user through the establishment of an environmental management system (EMS), in accordance with the requirements specified in ISO 14001:2015, *Environmental management systems — Requirements with guidance for use.*

There are many benefits to implementing an EMS. These include a potential for:

- improved control and management of emissions, effluents and wastes;
- avoidance and safe handling of hazardous or potentially polluting materials;
- reduction in generated wastes;
- energy efficiency improvements and cost savings;
- conservation of natural resources, including water, land and precious minerals;
- a comprehensive approach to satisfying legal and other requirements;
- operational efficiency and cost savings;
- pursuing environmental initiatives that are aligned with business priorities.

Other benefits can include increased profitability, better access to markets and improved relationships with stakeholders (e.g. customers, regulators, investors, insurers, neighbours). In some cases it may be of further value to organizations and their stakeholders to have the EMS assessed for conformity in order to demonstrate that it meets the requirements of ISO 14001:2015 [1)].

This guide is aimed at small and medium-sized enterprises (SMEs), both in developed, transitioning economy and developing countries, but can be used by an organization of any size. It provides an overview of the ISO 14001:2015 requirements and promotes the guidance offered in ISO 14004:2016, *Environmental management systems — General guidelines on implementation.* It seeks to provide practical help that assists organizations in establishing an EMS that can address its particular needs.

---

1) Section 10, *Demonstrating conformity*, describes the various options for assessing and demonstrating conformity to ISO 14001.

The guide has twelve sections, seven of which cover a particular stage in the EMS implementation process. Each of these sections provides an explanation of the relevant requirement(s), as well as guidance on how to get started, helpful implementation techniques, examples and a self-assessment checklist to verify conformance. Other sections provide useful information on the environmental management system model, guidance on integrating management systems, conformity assessment options and additional reference materials. A fictional case study is also included. The information included in this guide, however, should not be misconstrued as modifying or adding to the ISO 14001 requirements.

ISO 14001:2015 and other management system standards like ISO 9001:2015, *Quality management systems — Requirements*, are closely related. The structures are identical and several elements of these two management systems contain the same requirements. However, there are also significant technical differences. If an organization has either implemented or is already certified to any of the ISO management system standards such as ISO 9001, it should have in place many of the components common to most management systems, for example, context of the organization, competence, documented information, internal audit, nonconformity and corrective action, to name a few. The task for organizations wishing to adapt their management system to address environmental issues will be to expand the scope of these common components, and address the remaining environment-specific requirements in ISO 14001:2015.

Since this guide does not include the text of ISO 14001:2015, users are recommended to obtain a copy from their national standards body or from ISO, either directly via sales@iso.org or via the Internet from www.iso.org.

1 Pursuing an environmental management system

1.1 Environmental management

Societal expectations for sustainable development continue to evolve due to growing pressures on the environment from pollution, natural resource scarcities, and global warming. Governments responded with increasingly stringent legislation. Organizations answered back by controlling their environmental impacts, primarily through pollution prevention and managing their waste streams. More recently, emphasis and positive efforts toward conserving natural resources and reducing use of fossil fuels for energy generation have been made. But expectations continue to mount for organizations to be more proactive in protecting the environment from harm and degradation.

Environmental management encompasses an organizations' efforts to control its interaction with and effect on the environment in order to minimize adverse environmental impacts and leverage positive environmental impacts. It is a core responsibility of all organizations. Stakeholder groups exerting pressure and holding organizations accountable for their environmental footprint is increasing: these stakeholders include board of directors, insurance companies, investors, customers, community members, regulators, and employees. Nongovernmental organizations are extending these pressures, holding countries, sectors and indeed organizations accountable for reducing pollutants.

An environmental management system (EMS) is an approach used by organizations since the 1990s to manage its interactions with the environment in a planned and systematic way. It comprises a comprehensive set of processes that an organization uses to establish and execute its policies and objectives. These processes include organizational structure, roles and responsibilities, planning, operations and performance evaluation. When implemented in unison, this system of processes focuses on making incremental improvement over time.

ISO 14001 provides such a framework for continual improvement of environmental management. It incorporates proven techniques implemented worldwide, and is internationally accepted. It is a voluntary framework which an organization chooses freely to undertake. It does not contain product specifications nor does it set requirements for performance outcomes. Organizations of any size or type, whether located in a developed, transition or developing country, can set for themselves environmental performance goals and use ISO 14001 to achieve those goals.

Most organizations that have implemented ISO 14001 have attested to significant improvement in minimizing wastes, conserving energy and improving their compliance status. These gains have been accomplished while improving operational efficiency and saving money. The most critical success factors include:

- leadership and commitment by top management,
- selection of environmental initiatives that address significant environmental impacts and are aligned with other business priorities, particularly when they yield a competitive advantage, and
- developing a proactive environmental culture throughout all levels and functions of the organization.

A summary of the key components of an environmental management system based on ISO 14001:2015 is provided in Table 1. Once fully implemented, an organization is on its way to contributing to the "environmental pillar" of sustainability.

Table 1 — Key components of ISO 14001:2015

Key concepts in ISO 14001:2015		
PLAN	1.	Understand the organization and its context, including environmental conditions.
	2.	Understand the needs and expectations of interested parties and determine which of those the organization will comply with.
	3.	Determine the scope (i.e. boundaries and applicability) of the EMS.
	4.	Establish and implement the environmental management system.
	5.	Garner leadership commitment from top management.
	6.	Establish an environmental policy.
	7.	Assign responsibilities and authorities for relevant roles.
	8.	Identify environmental aspects and associated environmental impacts, considering the product/service life cycle.
	9.	Determine the applicability of its compliance obligations, and address them within the EMS.
	10.	Determine the priority risks and opportunities to the intended outcomes of the EMS, including those related to significant environmental aspects, compliance obligations and other issues and requirements.
	11.	Plan to take actions to address significant environmental aspects, compliance obligations and other priority risks and opportunities.
	12.	Plan how to integrate the actions into its business processes and how to evaluate the effectiveness of the actions.
	13.	Establish one or more environmental objectives and a plan to achieve them, including indicators for monitoring progress.
DO	1.	Provide the resources required to implement and maintain the EMS.
	2.	Determine the skills and knowledge needed for the EMS, then obtain the necessary competence, including any required training.
	3.	Raise awareness of the EMS.
	4.	Establish, implement and maintain the processes needed for internal and external communications.
	5.	Create, update and control documented information needed for EMS effectiveness and required by ISO 14001.
	6.	Plan, implement, and control operations and processes needed to meet EMS requirements.
	7.	Prepare for and respond to emergency situations.

Key concepts in ISO 14001:2015		
CHECK	1.	Monitor, measure, analyse, and evaluate environmental performance.
	2.	Evaluate adherence to compliance obligations.
	3.	Conduct periodic internal EMS audits.
	4.	Review the EMS to ensure continuing suitability, adequacy and effectiveness.
ACT	1.	Make improvements by taking action to achieve the EMS intended outcomes.
	2.	Take action to deal with nonconformity and prevent its recurrence.
	3.	Take action to continually improve the suitability, adequacy and effectiveness of the EMS, focusing on elements that enhance environmental performance.

1.2 The Plan-Do-Check-Act model and ISO 14001

Successful organizations have found they can consistently achieve their strategic goals if they manage their business in a systematic manner. The systematic methods used may vary in detail, but typically include the following elements:

- planning
- operating and controlling
- monitoring and reviewing
- acting to improve.

The Plan-Do-Check-Act (PDCA) model provides an ongoing, iterative process that leads to continual process improvement. It is a proven methodology that enables an organization to establish policy commitments and take actions in a systematic manner to realize those commitments. This model is the underlying approach used in ISO 14001. It is a process that can be applied to the EMS as a whole and to each of its individual elements to continually improve environmental performance.

Figure 1 depicts the PDCA model when applied to the overall EMS. For demonstration purposes, its application to the single element, Competence, is illustrated in Figure 2.

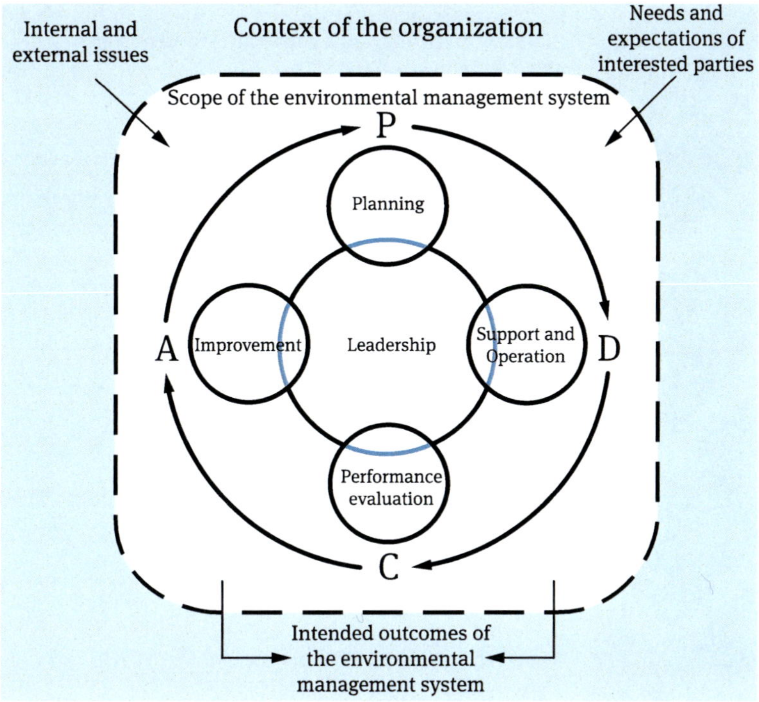

Figure 1 — PDCA model applied to an EMS

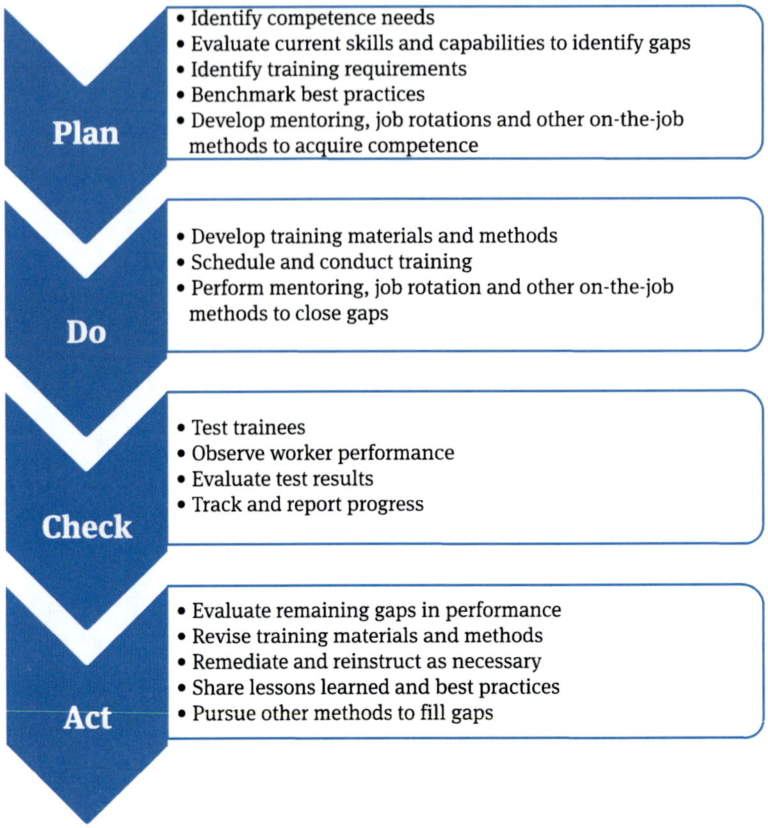

Figure 2 — **PDCA applied to a single EMS element, Competence**

1.3 What's new in the 2015 version of ISO 14001?

The launch of the ISO 14001 revision commenced in early 2012. Two drivers for the revision were:

1) New rules for ISO management system standards that required a common structure, a core set of requirements, and common terms and core definitions [2].

---

2) ISO/IEC Directives, Part 1, Consolidated ISO Supplement — Procedures specific to ISO, Annex SL.

2) Recommendations identified by TC 207/SC 1 members who studied the challenges associated with EMS implementation, including the evolving stakeholder expectations and advancements in management practices, since ISO 14001 was first published in 1996 [3].

In addition, user feedback from over 5000 organizations provided valuable input on the ISO 14001 revision [4]. The key changes are described below.

Strategic environmental management — A new requirement to understand the organization's context has been incorporated to identify and leverage strategic opportunities for the benefit of both the organization and the environment. Opportunities can arise from issues or changing circumstances related to the organization. Particular focus is on local, regional or global environmental conditions that can affect, or be affected by, the organization, and the associated relevant needs and expectations of interested parties (including regulatory requirements).

Risk-based thinking — Once the strategic issues (described above) are identified, the organization decides which ones are a priority to address in its environmental management system based on the importance to mitigate the adverse effects or value of leveraging the beneficial effects on the organization. The risk-based thinking related to the assessment of environmental aspects associated with an organization's activities, products and services are retained in the 2015 version of the standard.

Leadership — To ensure the success of the system, a new clause has been added that assigns specific responsibilities for those in leadership roles to promote environmental management within the organization.

Protecting the environment — Organizations are expected to make an environmental policy commitment to proactively protect the environment from harm and degradation. This commitment includes prevention of pollution initiatives, and can also include other areas that are relevant to its context,

---

3) ISO/TC 207/SC 1, Future challenges Study group N 9, Final report on the future challenges of EMS and ISO 14001, 2010.

4) TC 207/SC 1 N1016, ISO 14001 Continual Improvement Survey 2013 - Final Reports and Results, 2013.

such as sustainable resource use, climate change mitigation and adaptation, protection of biodiversity and ecosystems, etc.

Environmental performance – In addition to a new requirement to implement improvement opportunities that support achievement of intended outcomes, emphasis is placed on focusing continual improvement efforts on the management system elements that drive improved environmental performance, such as continually improving operational controls to reduce emissions, effluents and waste to target levels set by top management.

Life cycle thinking – In addition to managing environmental aspects associated with its procured goods and services and its production processes, organizations will need to consider if it can apply controls or influence to the management of environmental aspects associated with product design, transportation, use, end-of-life treatment or final disposal.

Communication – The development of a communications process with equal emphasis on external and internal communications has been added.

Evaluation of effectiveness – The need to evaluate the effectiveness of the EMS has been expanded from the requirement in 2004. Previously, only the effectiveness of corrective and preventive action and the effectiveness of the EMS (during management review) had to be evaluated. In the 2015 version, the following additional requirements are related to the effectiveness of the EMS:

- the leadership takes responsibility for the effectiveness of the EMS;
- reviewing the effectiveness of actions planned to address risks and opportunities;
- evaluating the effectiveness of competence training actions;
- ensuring staff understands their role in achieving the effectiveness of the EMS; and
- evaluating the effectiveness of the EMS.

In all cases, an organization can use the performance evaluation elements of ISO 14001 to evaluate effectiveness (i.e. monitoring and measuring, evaluation of compliance, internal audit and management review).

Intended outcomes — A new concept in the 2015 standard that represents the overall result that an EMS intends to achieve. The following three intended outcomes are specified, but organizations may identify additional ones:

- enhancement of environmental performance,
- fulfilment of compliance obligations, and
- achievement of environmental objectives.

Analysis and evaluation — Although previously implicit in the 2004 version, the 2015 version contains an explicit requirement to analyse and evaluate monitoring and measurement results, and to evaluate environmental performance. Further, there is increased emphasis on evaluating the effectiveness of specific actions taken and EMS effectiveness overall.

1.4 Drivers for a Small or Medium size Enterprise (SME)

An EMS provides value for all organizations but in particular it can help SMEs. Benefits can be summarized in the following categories:

1) Eliminates, minimizes and mitigates the environmental impacts associated with products and services offered by the organization.
2) Cuts costs and increases profits by preventing pollution and protecting the environment.
3) Creates a competitive advantage by providing products and services with lower environmental impact.
4) Drives growth by opening up new customers and export markets for products and services that meet ISO 14001 conformance requirements.
5) Enhances credibility with external interested parties and secures customer confidence.
6) Streamlines business processes and increases operational efficiency by focusing on continual improvement.
7) Helps the organization identify and comply with applicable legal and other requirements.

Each organization has the flexibility and autonomy to decide the level of detail and complexity of an EMS, the extent of documented information and the resources devoted to it. The decisions can be based on a number of factors, such as the scope of the system, the size of an organization and the nature of its activities, products and services. This is particularly important for smaller organizations.

1.5 Using this guide

This guide provides guidance on ISO 14001:2015 requirements and how to augment an organization's existing processes for managing its environmental responsibilities to conform to ISO 14001. It can be used by organizations of all sizes and types regardless of the nature of their activities. ISO 14001 can be implemented in whole or in part to systematically improve environment management practices.

This guide covers each of the ISO 14001 clauses in the standard. A short explanation of its importance is given. Key terms are defined. The requirements are explained in layman's terms, as much as possible, with examples, tools and templates to illustrate some implementation methods. Helpful hints and special considerations collected from experienced implementers are provided. Tips for building on existing processes are given. Documentation and communication requirements are highlighted, as well as important linkages to other elements. These linkages represent the other clauses that have related requirements. An overview of this guides' structure is provided below.

Pursuing an environmental management system	This section gives an overview of what an ISO 14001 environmental management system is and how it has benefited organizations, including SMEs. Suggestions on how to start the journey of implementing an EMS are given.

Understanding the organization's context **Setting the direction for the EMS** **Determining risks and opportunities** **Establishing EMS support processes** **Maintaining control of EMS processes** **Evaluating EMS performance** **Improving the EMS and environmental performance**	These sections provide guidance to help readers understand ISO 14001 requirements, together with examples and suggestions of how these requirements could be met.
Integrating other management systems	This section gives guidance and tips for implementing multiple management systems each having a different technical discipline, such as quality management, occupational safety and health, food safety or others.
Conformity assessment	This section gives the options available to organizations wishing to demonstrate conformity to ISO 14001. The four methods are explained and a detailed description of third party certification process is provided.
Case study	This section provides an EMS implementation case study for a small fictitious manufacturing organization.
Finding more information	This section provides references to sources of information which the reader might find useful.

1.6 Getting started

Whether setting out to implement an ISO 14001 conforming EMS for the first time, or upgrading an existing EMS to meet the 2015 revision requirements, some preliminary planning steps are recommended, see Figure 3.

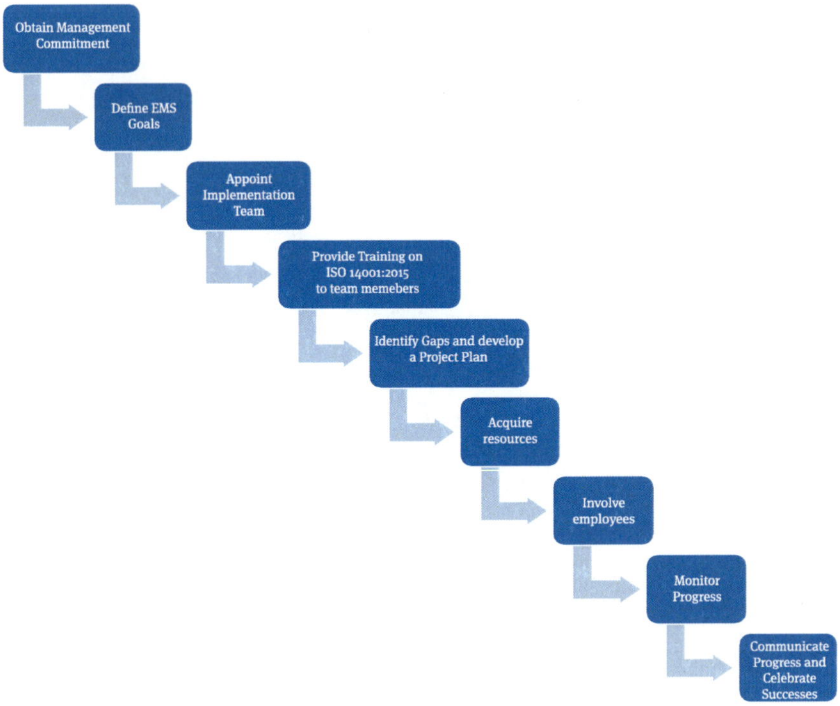

Figure 3 — Planning for an EMS

2 Understanding the organization's context

Before setting out to establish an EMS, a strategic understanding of the important factors that will affect the design and focus areas is needed. Each organization is unique and needs to identify what it plans to address and achieve so that the EMS brings value to the organization. ISO 14001 contains four clauses related to this effort:

- Organizational context
- Interested party needs and expectations
- Decision on the scope of the EMS
- Overarching EMS requirements

2.1 External and internal issues important to the EMS (ISO 14001:2015, Clause 4.1)

The organizational context is akin to an honest and objective situational analysis of the organization's environmental practices. It looks internally and externally at factors that concern the organization's environmental responsibilities and can, in part, affect its success. This situational analysis provides the "scene-setting" for determining issues that present risks and opportunities that need to be addressed in the EMS. It does not require a detailed analysis of these issues, but it does require a conceptual understanding of the potential adverse or beneficial effect these issues can have. It is important that the organization's leaders are involved in this analysis.

2.1.1 Getting started

2.1.1.1 Taking stock of your current state

Whether it is formal or informal, all organizations do some type of business planning and goal setting. They evaluate how the business is doing and what they need to do differently to increase customers or improve their operations. Whatever process is used, it can be expanded in scope to incorporate the evaluation of issues that pertain to environment management.

Sources of information

- Mission and vision statements
- Strategic plans
- Historical performance data
- Information on local and regional environmental conditions
- Legal requirements
- Customer requirements
- Benchmarking
- Best practices

2.1.1.2 Why is organizational context important?

Issues are important topics for the organization, problems for debate and discussion, or changing circumstances that affect the organization's ability to achieve the goals it sets for itself. These issues can include the organization's strengths, its weaknesses, or changes occurring either within its operations (e.g. introduction of new product lines or activities) or externally (e.g. new stakeholder requirements).

Some of these issues can affect the organization positively by presenting opportunities that can be leveraged, or negatively by posing threats that need to be addressed to prevent potential harm. In terms of environmental management, the issues that are most relevant are those that either help or hinder the organization's ability to achieve the strategic goals of its EMS. ISO 14001 refers to the strategic goals of an EMS as "intended outcomes", which include:

- improving its environmental performance,
- meeting its compliance obligations, and
- achieving its environmental objectives.

Organizations may set additional intended outcomes for its EMS, but these are the minimum required for any EMS.

The intent is to gain an objective picture of where the organization stands with respect to how it is managing its operations in balance with its environmental responsibilities. It starts with gathering facts and understanding relevant trends about external and internal pressures and success factors that have affected it in the past or likely to affect its future. Since this is an EMS, particular focus is on environmental conditions – how they can affect, or be affected by, an organization. However, this is not intended to be a voluminous and detailed data collection activity, but an opportunity to reflect on the "bigger picture" issues facing the organization.

By discussing and debating the relevance of these factors on the organization's goals and the intended outcomes of its EMS, the management team can identify which are likely to affect the organization, both long-term and within the current planning period. Focus on the factors that will contribute positively to the success of the business and to achieving the outcomes expected from the EMS.

Key linkages

- Interested parties
- Risks and opportunities
- Environmental aspects
- Compliance obligations
- Environmental objectives
- Management review
- Continual improvement

Environmental conditions

An "environmental condition" is the state or characteristic of the air, water, land, flora or fauna at a certain point in time. It can be described in terms of biodiversity, ecosystems, climate or other characteristics. It can be viewed locally, regionally, nationally or globally, over time or related to a specific event, such as a volcano eruption.

Organizations can affect the environmental condition through the emissions, wastes and wastewater that is generated by its operations, and by the raw materials that go into its products, aka its environmental aspects. For example, the wastewater discharges from plating operations can affect the water quality (pH, total suspended solids, and heavy metals) of a neighbouring waterway.

Alternatively, the existing environmental conditions can also affect an organization's operations, and in some cases this effect can be detrimental. For example, during heavy rainfall, eroded marshlands can cause flooding to businesses located on the seashore, potentially shutting down operations and causing damage to the facility infrastructure.

Assessments of the environmental condition are frequently the function of government agencies, non-governmental organizations, and scientific and research institutions rather than the function of an individual business. But this information can be used by businesses to

- identify and manage its significant environmental aspects;
- determine if local or regional environmental conditions pose a potential issue.

Since minimizing adverse impacts on the state of the environment is at the heart of ISO 14001, understanding its current condition and how the organization can impact it, either positively or negatively, is the most important factor in understanding the organization's context.

What should be considered?

The first step in understanding the organization's context is identifying issues that can affect achievement of its EMS intended outcomes. These issues will vary from one organization to the next, but it can start by considering the questions in Table 2 to determine which are most relevant.

Table 2 — Organizational context considerations

Table 2A — Internal considerations	Table 2B — External considerations
How do the organizational activities, products and services affect the local or regional environment (i.e. its environmental aspects)? Have they caused contamination in the past?	Are local or regional environmental conditions affecting the organization's operations, either in the short-term or long-term? Will availability of natural resources, or lack thereof, affect it?
How is the organization doing managing its environmental aspects (i.e. environmental performance)?	What is the organization's competitive position with regard to environmental performance? Are competitors' products more environmentally friendly or are their operations performed with less adverse environmental impact?
Are there any planned changes to its current activities, products and services, and will this affect, either positively or negatively, its environmental aspects?	Are there new technologies that can be employed to improve environmental performance (i.e. production equipment technology, environmental control technology, information technology, etc.)?
What is the current compliance status? Does it need improving?	Are there new or changing environmental regulations that will need to be addressed? If expanding to other countries, will there be new regulations to implement?
What is the business strategy, and how will it affect achieving the EMS intended outcomes?	Are customers imposing environmental requirements? Have stakeholders expressed concerns about the organization's environmental performance?
Are the necessary and sufficient resource capabilities available for the EMS (i.e. staffing levels, environmental competency, financial, infrastructure)?	Are economic conditions impacting EMS resources, either positively or negatively?
How will the management style, policies and organizational culture influence the EMS?	Are social, cultural or political conditions impacting the EMS?
Are employees engaged in achieving the EMS intended outcomes?	Are supply chain practices performed in an environmentally responsible manner?

Answers to these questions provide information that can help the organization successfully manage its environmental responsibilities. The internal considerations can provide insight on its strengths and weaknesses — what it does well and where it can improve. The external considerations may bring to light a potential opportunity with beneficial effect if pursued, or a threat with detrimental effect if not addressed. The tool "Strengths, Weaknesses, Opportunities and Threats analyses" (SWOT) can help the organization focus at the strategic level.

NOTE Although split into two clauses in ISO 14001:2015, organizational context includes an understanding of the needs, wants and expectations of interested parties (see Section 2.2).

2.1.1.3 What is relevant?

When performing the situational analysis of organizational context, there may be many factors that create issues — problems or circumstances — that pertain to the organization's current approach to environmental management. The reality may be, however, they cannot be tackled all at once.

ISO 14001 uses the phrase, "relevant to its purpose and that affects its ability to achieve the intended outcomes of its EMS" to help an organization determine the relevant issues. The organization focuses on issues that are relevant to its mission and can affect its ability to enhance its environmental performance, fulfil its compliance obligations, and achieve its environmental objectives.

The organization retains the authority and accountability to decide the number of issues it addresses within its EMS: it does not have to address everything at once. When the organization begins planning its EMS (in Section 6.1), it will establish a process for applying a risk-based approach to these issues in order to identify those that pose the most adverse effect and those that can be leveraged for the most benefit. Through that process, the organization can select and focus on the most pressing issues. Then, through the processes of management review and continual improvement, additional issues can be addressed over time.

ISO 14001 does not require a detailed analysis or documentation of these issues but it may be useful to keep a record of them for future follow-up. At this point, a conceptual understanding of the relevant issues is all that is needed. A more in-depth analysis of the adverse or beneficial effect, which ones need to be addressed and the actions that will be planned to address them in the EMS is determined in Section 6.1.

2.1.2 Helpful hints

Context changes over time; things that are not important initially, may eventually become so. Using a team approach, led by top management, can be helpful in identifying the relevant issues for the business.

Be alert to implicit assumptions that are being made to verify they are valid and remain so in the future.

Smaller organizations, in particular, need to be realistic about the resources they have available and their ability to address the multitude of issues that may be identified. It is best to start with a small number of the most important issues to ensure that the gains realized can be sustained. Over time, additional issues can be tackled.

When looking at the external factors, the focus should be on the associated local and regional implications, although, in some cases, the national or international implications may also be relevant.

Recent societal trends indicate that businesses are expected to exercise better control over the environmental impacts associated with their supply chain.

With the increase in social networking, managing reputational risks will become more important.

2.1.3 Check conformity

Requirement	Y	N	Follow-up actions
Identify relevant internal issues that are important, problematic or changing.			
Identify relevant external issues that are important, problematic or changing.			
Determine if environmental conditions can affect, either positively or negatively, the organization.			
Determine if environmental conditions can be affected, either positively or negatively, by the organization's activities, products or services.			
Determine if these factors pose any other issue(s) that can affect, either positively or negatively, achievement of the EMS intended outcomes.			

2.2 Needs and expectations of interested parties (ISO 14001, Clause 4.2)

It goes without saying that an organization's success is based on doing what has to be done, first and foremost, and also doing what can be done to satisfy additional stakeholder expectations. The key is having a clear idea of which stakeholder groups are relevant, and then deciding which needs and expectations have to be addressed and which ones make sense to address. Understanding the views of stakeholders or interested parties (whichever term used) is one component of the organization's context and provides input when determining issues that are important to address in the EMS.

2.2.1 Getting started

Depending on the organization, there will be an array of interested parties that are important to its success. ISO 14001 defines interested parties as persons or organizations that can affect, be affected by or perceive itself to be affected by a decision or activity by your organization. Finding out what they need and expect can seem like a daunting task. Figuring out the right ones to address within the EMS, besides mandatory requirements the organization has to comply with (particularly legal requirements), requires a balance between competing, and often conflicting, requests and the resource capability to satisfy them. As with issues that organizations face when understanding its context, each organization retains the authority and accountability to decide which interested party requirements it addresses within its EMS: it may not be able to address everything at once. The ones that cannot be addressed initially can be reconsidered later when the organization performs its management reviews and sets new priorities for improvement.

ISO 14001 does not require a detailed analysis or documentation of these interested party requirements. At this point, a conceptual understanding of what the needs and expectations are is all that is needed so that a decision by management can be made on which interested party requirements it will address in its EMS. A more in depth analysis of these requirements is determined in Section 6.1.

Interested party requirement vs organizational requirement

Some clarification of ISO terms is necessary to avoid confusing an interested party requirement from an organization's requirement. ISO describes a "requirement" as a need or expectation that is stated, generally implied (i.e. custom or common practice) or obligatory (i.e. law or regulation). Interested parties are persons or organizations that can either affect or be affected by a decision your organization makes or by an activity it performs. Sometimes the effect of these decisions or activities have are debatable; the effect may be perceived but may not be supported by fact. In either case, the important distinction is that these represent the interested party needs and expectations, not necessarily the organization's requirements. Each organization determines who the relevant interested parties are and which of their needs and expectations are relevant to the organization. Once relevance is determined, then the organization can decide which ones it has to comply with and which ones it chooses to voluntarily comply with. ISO 14001 refers to this subset of requirements as the organization's compliance obligations.

2.2.1.1 Taking stock of your current state

There are several channels available to learn about interested party requirements related to environmental management. Legal requirements are conveyed through governmental authorities. Official information about these requirements are found in laws and regulations and can often be found online via the Internet. In some cases, specific requirements are communicated to organizations through permits to operate.

Other requirements might be communicated formally (i.e. contracts, charters, industry-related protocols and written communiques) or informally (i.e. face-to-face meetings and phone calls).

What methods are used to gather and compile information on the needs and expectations of interested parties? Are they proactive or reactive? If they are only reactive, the organization may be incorrectly assuming "no news is good news".

2.2.1.2 Internal interested parties

The most important internal stakeholders are employees. They are relied upon to implement the day-to-day actions that determine the success of the organization's EMS, so seeking their input and involvement will convey commitment and nurture a strong environmental ethic. Millennials want to make a difference, and working for an organization that makes a positive impact on the environment is often one of their expectations when evaluating employers. Where the labour force is unionized, it is important to engage the worker representative(s) to gain their support. Seek out their views on environmental concerns regarding work activities, products and services during meetings or via surveys. Follow-up on inquiries, requests, complaints or suggestions made by employees to learn more about their expectations.

In cases where there is a parent organization, its needs and expectations with respect to the environment are often contained in environmental policies or procedures.

The organization may have other important stakeholders — a board of directors, advisory board, investors or donors. They expect the organization's long-term viability, its reputation, and their investment to be protected. Environmental incidents, regulatory noncompliance or irresponsible actions of suppliers are potential issues which could concern them because these issues can jeopardize the business. A conversation with these stakeholder groups is usually sufficient to determine their expectations.

Table 3 — Examples of interested party

Internal	External
Employees	Customers
Managers	Neighbouring community members
Investors	Suppliers
Board of directors	Regulators
	Trade associations

2.2.1.3 External interested parties

Start by looking at what is formally stated in customer or supplier contracts, laws and permits, or terms in voluntary agreements.

Customers typically specify compliance with applicable laws in the contract terms and conditions. They may send surveys asking about material use or requests to participate in voluntary reporting schemes. When sending out "requests for bid", they may ask for cost saving measures which can be addressed through, amongst others, waste reduction initiatives.

Providers of waste management or public water treatment facilities may set requirements to which the organization must comply. Tenancy agreements may stipulate requirements set by the building owners.

Local, state, national or provincial regulators govern what can be emitted, stored or disposed. Specific limits may have been set on the use of particular hazardous materials, for example, in a facility permit to operate.

There may be environmental criteria found in voluntary agreements that the organization has entered into, such as trade association membership, loan agreements or reporting schemes.

Next, review inquiries, requests, complaints or suggestions made by community members to understand their views and environmental concerns regarding work activities, products and services.

It is not required to seek out input from interested parties who have not contacted the organization directly to express their concerns. However, as resources allow, an organization may choose to incorporate proactive measures to seek input from stakeholder groups using their communication process.

Key linkages

- Organizational context
- Scope of the EMS
- Risks and opportunities
- Compliance obligations
- Communication
- Management review

2.2.2 Helpful hints

Review formal requirements first, including those in contracts, agreements, laws and operating permits. Next, review verbal inquiries or requests. Then, seek input and feedback through proactive methods.

The task of understanding needs and expectations can be done separately or in combination with the more detailed analyses of the organization's compliance obligations, required by Clause 6.1.3 of ISO 14001 (see Section 4.2).

Smaller organizations in particular need to be realistic about the resources they have available and their ability to address needs and expectations expressed by interested parties. It is best to start with legal requirements and customer requirements to ensure the organization can sustain compliance with them. Over time, additional voluntary requirements can be added.

Interested parties can change over time so it is important to re-evaluate them during the management review process.

2.2.3 Check conformity

Requirement	Y	N	Follow-up actions
Identify relevant internal interested parties.			
• Determine the environmental requirements they expect of the organization, its operations or its products.			
Identify relevant external interested parties.			
• Determine the environmental requirements they expect of the organization, its operations or its products.			
Decide which environmental requirements the organization has to comply with.			
Decide which voluntary environmental requirements, if any, the organization chooses to comply with.			

2.3 Scope of the EMS (ISO 14001, Clause 4.3)

2.3.1 Getting started

It is important to clearly articulate the physical and organizational boundaries of the EMS. This defines where the EMS is applicable and will be implemented. The scope of an EMS can include the whole organization, or specific and identified functions or sections of the organization. If the organization makes a statement that it conforms to ISO 14001, then it must make the scope statement available so that interested parties clearly understand what parts of the organization conform.

2.3.1.1 Taking stock of your current state

Look at the current environmental programmes. If there are particular organizational units, work areas, or functions that it applies to, then it makes sense to include those work areas at a minimum. If there are known gaps in the environmental programme, areas where it is known more work needs to be done to meet requirements, then it makes sense to include those work areas also.

Sphere of control is an important consideration when defining the scope of an EMS. The top manager(s) must have the authority to make the policy commitments and implement an EMS within the functional areas or facilities that are defined within the scope.

2.3.1.2 Determining the scope of the EMS

The first step in determining the scope is looking at the physical boundaries of the organization and the functional units included, then at the activities products and services conducted within. Top management should decide the extent to which they want to implement their EMS. The important inputs for making the decision include the issues that were identified during the analysis of the organizational context (Section 2.1), and the compliance obligations that were identified (Section 2.2).

Once the scope is defined, operations that have or can have significant environmental aspects or applicable compliance obligations cannot be excluded. A common pitfall is not including the products and services that the organization produces. This also includes the goods and services that the organization uses in its operations that it procures from its suppliers. The scope needs to accurately reflect the facilities and functions included to avoid misleading interested parties and potentially undermining the credibility of the organization's EMS.

NOTE The organization's EMS scope is required to be documented and kept up-to-date.

There are several methods for documenting the scope of an EMS, including written descriptions, or diagrams. Whatever approach is used, the key components are: i) identification of the activities involved, ii) the products and services that result, and iii) the location where it occurs.

Key linkages

- Organizational context
- Interested parties
- EMS
- Environmental aspects

2.3.2 Helpful hints

Some organizations start small, and implement an EMS in a phased approach so they can test their EMS processes before implementing them across the whole organization.

Once tested, implementation of the EMS can be expanded across the organization to ensure consistent implementation of environmental practices and to include all functions involved in the product life cycle.

If the EMS scope is limited to a subset of the organization, "top management" refers to those managers who are responsible for that part.

The scope should be re-evaluated when the business expands, adds new facilities or divests portions of its operations.

Communicating

The scope of the organization's EMS needs to be available to interested parties. Options include providing a written description when inquiries are received, including it on the company website, or posting it in the lobby where visitors can see it. If the company undergoes third party certification, the scope description is included on the certificate, so posting the certificate in the lobby can fulfil the requirement to make the scope available to interested parties.

2.3.3 Check conformity

Requirement	Y	N	Follow-up actions
Determine the boundaries and applicability of the EMS (i.e. scope of the EMS).			
Document the scope.			
Make the scope available to interested parties.			

2.4 Environmental management system (ISO 14001, Clause 4.4)

2.4.1 Getting started

This clause of ISO 14001 specifies a general, overarching requirement for an organization to systematically manage its environmental responsibilities in a manner that enhances its environmental performance, fulfils its compliance obligations and achieves the environmental objectives it sets for itself.

An environmental management system encompasses the organizational elements and set of processes an organization uses to establish and deploy its environmental policy and environmental objectives. ISO 14001 stipulates the set of required elements needed for an environmental management system in each of its clauses.

An organization satisfies this general requirement by conforming to the provisions specified in all clauses of ISO 14001. Not only must an organization establish an EMS, they need to implement, maintain and continually improve it, including any processes necessary to ensure the intended outcomes can be achieved.

2.4.1.1 Taking stock of your current state

All organizations have at least some of the EMS elements in place. By performing an initial review of current management practices, gaps can be identified. Collect information on existing processes to see what is already in place, what processes can be enhanced to meet ISO requirements, and lastly, what additional information and processes do not yet exist and will have to be established.

Don't limit the review to the environmental department. Other functions, such as human resources, procurement, quality, safety and others, may be implementing processes and generating information similar to that required by ISO 14001. By incorporating the environmental components into existing processes, the organization can integrate the EMS requirements into its business processes.

2.4.1.2 Integration with business processes

One pitfall organizations sometimes make is designing their EMS as a stand-alone system, independent of its routine, standard management practices. Initially, this may meet ISO requirements, but over time actual practice will deviate because normal management practices prevail. For this reason, it is important to integrate the EMS requirements into the standard management practices and business processes.

For example, if an organization conducts an annual strategic planning process to review its market position, customer needs and business performance, then it is more effective to incorporate the assessment of organizational context, interested party needs and expectations, and environmental performance into that process. By doing so, environmental issues can be evaluated in light of the organization's strategy, and environmental initiatives can be aligned with other business priorities. Further, a process that is already well established within the organization is leveraged.

Key linkages

- Organizational context
- Compliance obligations
- Continual improvement

Plus, see all other sections of this guide which describe the individual requirements of an EMS.

2.4.2 Helpful hints

Each organization can choose the way it will satisfy ISO 14001 requirements so that it best suits the organization's management style and practices. Likewise, it has the autonomy to decide the extent to which it will integrate its EMS into its business processes.

NOTE The documented information that is required for the EMS is specified in the respective sections of this guide.

When starting out, focus on areas where benefit will be reaped. Eliminating or reducing environmental aspects will protect the environment, prevent pollution and improve the organization's environmental performance. It will have the added benefit of potentially reducing the number of regulations that apply to its operations.

When establishing EMS processes consider how operational efficiency can be improved, administrative burden reduced and cost savings realized.

If the EMS is implemented within a specific operating unit, policies and procedures developed by other parts of the organization can be used to meet requirements as long as they are applicable to that specific unit.

2.4.3 Check conformity

Requirement	Y	N	Follow-up actions
Establish an EMS in conformance with ISO 14001			
• Consider the issues and compliance obligations identified (Sections 2.1 and 2.2, respectively)			
• Apply to the activities products and services that are within the scope of the EMS			
• Include the processes needed and their interactions.			
Implement the EMS.			
Maintain the EMS.			
Continually improve the EMS.			

3 Setting the direction of the EMS

Top management has a prominent and important role in setting the expectations for the EMS and inspiring the workforce to fulfil those expectations. ISO 14001, identifies three clauses for top management to assist in this effort:

- Leadership
- Setting environmental policy
- Assigning roles, responsibilities and authorities.

3.1 Leadership and commitment (ISO 14001, Clause 5.1)

The experience of many organizations has shown that visible support, involvement and commitment by top management are critical success factors when implementing an EMS. It has also been shown that an EMS aligned with the business strategy can help an organization achieve its business goals. Top managers can appreciate their role as "Leaders" in the organization and their ability to inspire others. They have a responsibility to build an organizational culture that will guide worker behaviour both on and off the job. To realize these synergies, top management needs to understand their ownership role in the EMS.

3.1.1 Getting started

3.1.1.1 Taking stock of your current state

What methods do top managers use to communicate new initiatives? How do they actively garner support? Explore how these methods can be used and enhanced to foster a proactive environmental culture within the organization.

3.1.1.2 Top management as leaders

Top management is the person or group of people who direct and control an organization at the highest level. They have the power to delegate authority and allocate resources. For some organizations this may include the owner or partners, the facility manager and a few key people who report directly to them. For others, it may expand to include board of directors, investors, or, in some cases, the top management of contractors. If the scope of the EMS is limited to only a part of the organization, then top management refers to those who direct and control that part.

These people have a direct role in the EMS. There are particular actions they must do, or direct to be done, in order to lead others in EMS implementation and demonstrate their personal commitment (see Clauses 5.1, 5.2, 5.3, and 9.3). Top management can delegate some of the tasks and activities involved, which are identified in ISO 14001 with the word "ensure", but in the end, top management is accountable for the results.

Top management

- Takes accountability for EMS effectiveness
- Communicates importance of EMS and conformance to its requirements
- Directs/supports others to contribute to EMS effectiveness
- Promotes continual improvement
- Supports others in demonstrating their leadership

Top management ensures

- Compatibility of environmental policy and objectives with the organization's context and its strategic direction
- EMS integration into the business processes
- Resource availability
- Achievement of intended outcomes

Communicating

- Leaders communicate the importance of effective environmental management.
- Leaders communicate the importance of conforming to the EMS requirements.

3.1.1.3 Demonstrating leadership

There are a multitude of ways top management can demonstrate that it has fulfilled these requirements.

- Accountability can be demonstrated by assigning and assessing environmental goals for top managers and others during the performance appraisal process.
- Records of talking points during staff meetings can provide proof of discussions on the value and benefits of the EMS and adherence to its requirements.
- Corrective action assignments are ways managers direct others to take action to make the EMS more effective.
- Management review meeting reports show top managements' role in monitoring progress on environmental objectives and evaluating performance.
- Understanding how relevant internal and external issues affect the organization's business goals and environmental responsibilities, and taking that into account when setting the environmental policy commitments and environmental objectives.

- Making sure functional units are aligned on the EMS priorities and are performing their EMS responsibilities.
- Maintaining focus on improving environmental performance, satisfying compliance obligations and achieving environmental objectives when conflicts with other business priorities occur.
- Providing the necessary resources to plan and fulfil the EMS requirements, including personnel time, training, financial support, and necessary equipment and infrastructure.
- Setting and communicating priorities for improvement based on analysis of environmental data, audit recommendations and management reviews.
- During EMS audits, top management describe their involvement in the above activities to auditors.
- Leading by example through personal action.
- Employees provide the ultimate proof when they attest to managements' involvement and support.

Key linkages

- Roles, responsibilities, authorities
- Resources
- Communication
- Management review
- Improvement

3.1.2 Helpful hints [5]

Establish a vision that captures the level of performance expected, then use passion and persuasion to promote the EMS and to enlist others in making the vision real for the organization.

---

5) Excerpt from "The Five Practices of Exemplary Leadership® Model", The Leadership Challenge, a Wiley brand; http://www.leadershipchallenge.com/About-section-Our-Approach.aspx.

Challenge the *status quo*. Encourage innovative ways to improve the organization's performance. With the team members, identify and promote proactive environmental management initiatives that can yield business benefits.

Set interim goals so that people can achieve small wins as they work toward larger environmental objectives. Unravel bureaucracy when it impedes action. Provide guidance and direction when conflicts occur or the next steps are unclear. Treat mistakes and failures as learning opportunities.

Foster collaboration and team spirit to build an environmental ethic among employees in the organization.

Share information on key environmental issues and performance indicators at regular management meetings.

Recognize individual contributions, reward their efforts and celebrate successes along the way.

3.1.3 Check conformity

Requirement	Y	N	Follow-up actions
Organizational leaders:			
• are accountable for EMS effectiveness			
• communicate the importance of responsible environmental management and adhering to EMS requirements			
• direct and support others to participate in EMS activities and do their part in achieving the planned results			
• promote continual improvement of the EMS and environmental performance			
• assist and support others to demonstrate environmental leadership in their work activities.			
Top managers ensure environmental policy commitments and objectives:			
• reflect the most important internal and external issues that pertain to the EMS			

Requirement	Y	N	Follow-up actions
• are aligned with and support the business strategy.			
Top managers facilitate the integration of EMS requirements into the core business processes.			
Top managers ensure its EMS improves environmental performance, meets compliance obligations and achieves environmental objectives.			
Top managers ensure sufficient resources for the EMS are available.			

3.2 Environmental policy (ISO 14001, Clause 5.2)

An environmental policy is the mechanism for top management to formally state their commitments to the environment. It sets the direction for the organization and describes the expected results from systematically managing its environmental aspects.

At a minimum, three commitments are required:

- protection of the environment,
- fulfilment of compliance obligations, and
- continual improvement of the EMS to enhance environmental performance.

The environmental policy commitments are the foundation for the EMS. The work practices put in place enable the organization to operate in a manner consistent with these commitments. Environmental objectives are set to drive improvement where needed to better align with these commitments. The EMS is designed to achieve the environmental policy commitments over time.

It is important to share the organization's commitments with its employees, contractors and others who work on behalf of the organization so they understand top managements' expectations and can perform their work in a manner that contributes to achieving those results. Making the environmental policy publicly available, or available upon request, provides assurance to community members and customers that the organization is doing its part to protect the environment.

Communicating the environmental policy

- Posters in the workplace
- Training classes
- Employee meetings
- Print on badges
- Business stationary
- E-mail signature statement
- Websites
- Annual reports

3.2.1 Getting started

3.2.1.1 Taking stock of your current state

What policies or environmental commitments does the organization already have? They may not be captured in an "environmental policy" *per se*, but there may be similar commitments in its ethics and compliance, quality or other organizational policies.

Take into consideration the organization's context, interested party expectations, and the scope of the EMS to be sure the commitments are relevant to the environmental impacts of its activities, products and services, and reflect the environmental challenges the organization faces.

3.2.1.2 Policy commitments

Three policy commitments are required by ISO 14001. Organizations may tailor their commitments so they are meaningful, achievable and stretch its capabilities. Top management owns these commitments and will need to provide the support necessary to demonstrate that it is living by its policy commitments.

By addressing the significant environmental aspects of the organization's activities, products and services, the environmental policy commitments selected can add value to the surrounding community and yield benefits to the organization in terms of strengthened relationships with stakeholders.

Protection of the environment

Protecting the environment starts with preventing pollution, harm and degradation. Recycling wastes, minimizing air emissions and substituting less hazardous chemicals are normally the first steps organizations take to preventing pollution. The approach selected should follow a hierarchy, giving preference to preventing pollution at the source, then recycling or reuse, then treating or controlling the pollutant.

ISO 14001 defines prevention of pollution as the use of processes, practices, techniques, materials, products, services or energy to avoid reduce or control (separately or in combination) the creation, emission or discharge of any type of pollutant or waste, in order to reduce adverse environmental impacts. Prevention of pollution can include:

- Source reduction or elimination
- Process, product or service changes
- Efficient use of resources
- Material and energy substitution
- Reuse
- Recovery
- Recycling
- Reclamation, or
- Treatment.

Beyond pollution prevention, there may be other commitments that are relevant to an organization's context worth pursuing. Curtailing energy consumption from fossil fuels can reduce CO_2, reduces operating costs and mitigates climate change. Limiting land development protects and conserves habitats, wildlife and the ecosystem. Select commitments that can make a beneficial impact in the surrounding locality or region. For example, water conservation may be appropriate for an organization located in an arid climate where water is scarce. Even more important for some organizations may be focusing on the design and development of its products and services that minimizes negative environmental impacts and maximizes positive benefits at each of the life cycle stages.

Incorporating policies such as zero waste to landfill, carbon neutrality or adoption of sustainable development principles are ways an organization can demonstrate their commitment to protecting the environment.

Fulfilment of compliance obligations

A responsible organization is expected to consistently adhere to its legal or other mandatory requirements as well as any requirements associated with voluntary obligations it agrees to undertake. Some organizations may not be in compliance with particular regulations, or may need to work on consistency in meeting the requirements. Others operate well within their regulatory limits so may choose to go beyond compliance by self-imposing more stringent criteria for their emissions or effluents.

NOTE The environmental policy is required to be documented and kept up-to-date.

Continual improvement

This concept is the underlying principle of a management system, and is addressed specifically in Clause 10 of ISO 14001. Although improving any element of the EMS adds value, a commitment to improving the EMS should focus on the elements that drive environmental performance improvement. Attention to those most directly related to environmental aspects of the organization's activities, products and services will be most effective in enhancing environmental performance. For example,

- Environmental aspects identification process
- Environmental objectives and planning to achieve them
- Operational planning and control processes
- Management review.

Further, ISO 14001 contains a general requirement for organizations to take action on identified improvement opportunities to achieve the intended outcomes of its EMS (see Section 8.2).

3.2.1.3 Implementation and maintenance

Setting the policy commitments is just the start; these commitments are then reflected in the processes and requirements established within the various elements of the EMS (see Key linkages).

The organization periodically assesses it progress toward meeting its policy commitments (i.e. during the management review process), making changes or pursuing additional opportunities as needed.

Key linkages

- Organizational context
- Interested parties
- Intended outcomes
- Leadership
- Environmental aspects
- Compliance obligations
- Environmental objectives
- Awareness
- Communication
- Operational control
- Management review
- Continual improvement

3.2.2 Helpful hints

Keep the policy simple so that it is easily understood by employees. They need to be aware of how it applies to their work and what they do to help achieve the commitments.

Often an organization will develop a draft environmental policy as a starting point. As more information becomes available about how it impacts the environment and how it plans to control these impacts, it can be modified over time.

Create a stand-alone environmental policy document, or consider integrating environmental commitments with other organizational policies such as occupational health and safety, quality management or business ethics.

Since the commitments in an environmental policy can evolve over time, communicate it in a way that makes it easy to update. Many organizations invest in posters, translations or banners that causes them to hesitate changing the policy. Framed policies can be easily changed.

3.2.3 Check conformity

Requirement	Y	N	Follow-up actions
An environmental policy is established that is:			
• relevant to the organization's purpose and context			
• reflects the type, magnitude and environmental impacts of activities, products and services within the scope of the EMS			
• used to set environmental objectives.			
The environmental policy includes the following commitments:			
• to protect the environment addressing prevention of pollution, harm and degradation			
• to meet its compliance obligations			
• to continually improve the EMS in order to improve environmental performance.			
The environmental policy is documented.			
The environmental policy is communicated internally.			
The environmental policy is available to external parties.			
The environmental policy is implemented.			
The environmental policy is maintained.			

3.3 Roles, responsibilities and authorities (ISO 14001, Clause 5.3)

Each person working for or on behalf of the organization has a role in the EMS; that is, to do their part in supporting the environmental policy commitments by adhering to EMS requirements. Some have a more extensive role implementing particular elements of the system within their work activities. Others will have the role of overall coordination of EMS implementation and maintenance. In all cases, top management needs to assign and communicate the role(s), responsibilities and authorities for operating the EMS and delivering results.

3.3.1 Getting started

3.3.1.1 Taking stock of your current state

Start by examining the current organizational structure. Identify the current roles that are responsible for environmental management. Consider the operations with significant environmental aspects and activities needed to ensure compliance. Identify who needs to be involved to ensure these operations are controlled and activities are performed.

Since resources can be limited, find opportunities to integrate EMS responsibilities within existing functions of the organization, such as manufacturing operations, facilities management, purchasing, and human resources, for example. If other existing management systems are in place, such as quality management, occupational health and safety or food safety, synergies may exist where there are similar roles and responsibilities. Not only will this enhance ownership of environmental management across the organization, but will create efficiencies.

Assign EMS roles, responsibilities and authorities with the same degree of formality that operational roles, responsibilities and authorities are assigned. Ensure persons assigned are competent to perform their EMS roles and have the commensurate resources to fulfil their responsibilities.

Communicating roles, responsibilities and authorities

- Face-to-face meeting
- EMS responsibility matrix
- Job descriptions
- EMS organization chart

3.3.1.2 Key EMS responsibilities

In addition to the leadership roles and responsibilities described in Section 5.1 of this guide, there are other tasks that need to be assigned within the organization. There is no "one size fits all" formula for determining which role should be assigned which responsibility. This is determined on a case-by-case basis taking into account the expertise of staff members, their availability and alignment with their job responsibility. Some tasks will benefit from a cross-functional team approach, particularly when determining the organization's context, identifying significant environmental aspects and other organizational risks and opportunities, and when setting environmental objectives. Teams should include the varied perspectives of management, supervisors and shop-level employees.

Key EMS responsibilities need to be assigned within the organization. Although not an exhaustive list, the template shown in Figure 4 can be tailored and used to identify the primary and secondary roles for performing, supporting and being accountable for each of the EMS responsibilities.

Figure 4 — "RACI" TEMPLATE

Concept: "Everybody's job is nobody's job!" Clearly assigned roles and responsibilities ensure successful execution.

Objective: Identify key roles and responsibilities for a set of activities.

Instructions: For each responsibility, enter the letters R, A, C, I under the respective role using the following guidelines, then communicate the responsibilities to the respective roles and gain agreement:

"R" — Responsible: the one assigned a task or action;

"A" — Accountable: the one ultimately responsible for the decision or action taken;

"C" — Consult: advisor, provides suggestions but not part of the decision making process;

"I" — Inform: those who need to be notified or communicated with.

EMS responsibility	Organizational role								
	Sr Manager	Ops Manager	Ops Supervisor	HR	Maintenance	Environment	Engineering	Finance	Employees
Planning and coordinating EMS implementation and maintenance.									
Understanding organizational context issues and interested party requirements.									
Determining risks and opportunities that need to be addressed and associated action planning related to:									
• environmental aspects,									
• compliance obligations, and									
• other issues and requirements.									

EMS responsibility	Organizational role								
	Sr Manager	Ops Manager	Ops Super-visor	HR	Mainte-nance	Environ-ment	Engi-neering	Finance	Employ-ees
Understanding applicable legal requirements and other voluntary obligations.									
Establishing policy commitments.									
Setting environmental objectives and associated action planning.									
Assessing competency requirements, including training needs.									
Establishing and implementing internal and external EMS communications process.									
Creating, updating and controlling documented information.									
Establishing operating criteria and controlling facilities, equipment and operational processes.									
Establishing controls to ensure environmental requirements are addressed in the design and development of products.									
Establishing and communicating environmental requirements to contractors and suppliers.									
Communicating environmental impacts during use, transport and treatment/final disposal of products.									
Preparing and responding to emergency situations.									
Monitoring/measuring and analysing environmental performance.									

EMS responsibility	Organizational role								
	Sr Manager	Ops Manager	Ops Super-visor	HR	Mainte-nance	Environ-ment	Engi-neering	Finance	Employ-ees
Reporting environmental performance to top management.									
Evaluating the organization's compliance status.									
Developing an audit program and conducting internal audits.									
Identifying and correcting nonconformities.									
Identifying and implementing continual improvement opportunities.									

3.3.2 Helpful hints

The responsibility for overall coordination of EMS implementation may be assigned to a single person or a team. Critical to success is that they be knowledgeable, an advocate for the EMS, and have easy access to top management.

Informal structures typically exist in small and medium size organizations whereby workers "wear multiple hats", leaving many of these roles and responsibilities resting on a few key personnel. It may be beneficial to spread the EMS responsibilities among other staff members. It may provide an opportunity to develop young staff with a strong environmental ethic who are interested in taking on a leadership role in the EMS program.

Embedding the EMS responsibilities into existing Human Resource (HR) department files (such as job descriptions and employee performance goals) improves integration with the HR department business processes.

In some cases technical environmental expertise may not be available within small and medium size organizations, therefore external help may be needed. For example, if the organization does not have staff with environmental compliance expertise, it may need assistance in gaining detailed knowledge of legal requirements and how they apply to the operations. Technical information on available emissions and effluent control technologies may be needed. It may be prudent to have an external expert evaluate the organization's compliance status to ensure a comprehensive review is conducted. Help for these functions may be available from industry associations, universities, regulatory agencies or consultants.

Key linkages

- Leadership
- Resources
- Competence
- Awareness
- Communication

3.3.3 Check conformity

Requirement	Y	N	Follow-up actions
Assign relevant EMS roles, responsibilities and authorities.			
Assign responsibility and authority for:			
• ensuring the EMS conforms to ISO 14001 requirements			
• reporting EMS and environmental performance to top management			
Communicate relevant EMS roles, responsibilities and authorities.			

4 Determining the risks and opportunities that need to be addressed

Determining what's important to the success of the EMS, and addressing those things that can jeopardize its success, goes on in the Plan phase of the Plan-Do-Check-Act model. ISO 14001 identifies five important areas to focus on:

- Other risks and opportunities
- Significant environmental aspects
- Compliance obligations
- Planning action
- Environmental objectives

4.1 Overview

Organizational context, including the needs and expectations of interested parties, has the potential to pose adverse effects, as well as beneficial effects to the environment, the organization or both.

When planning an EMS, the key is to identify and focus the organization's resources on mitigating adverse effects and pursuing beneficial effects. Determining which are the most important, however, requires a prioritization process that reflects the importance of

- assuring the EMS can achieve its intended outcomes (including enhancing environmental performance, fulfiling compliance obligations and achieving environmental objectives),

- preventing or reducing adverse environmental impacts and undesired effects to the organization, and
- achieving continual improvement.

Once the priorities have been identified, the organization plans how it will address them within its EMS.

Since this is an environmental management system, the key risks and opportunities are those associated with environment aspects and compliance obligations. However there may be others that warrant being addressed by the organization.

> There is always an amount of uncertainty with issues that pose risks and opportunities. How likely is it that the negative result will occur? Will the effect be as bad or good as initially envisioned? Acknowledging that there is uncertainty, management needs to determine the strategically important risks and opportunities that support the EMS intended outcomes.

Emergency situations have the potential to cause an adverse environmental impact, and therefore need to be considered during the determination of environmental aspects (see Section 4.1). Further, they may cause adverse effects on the organization, so should be considered when determining risks and opportunities. Therefore it is important to identify reasonably foreseeable emergency situations when planning the EMS. Additional guidance on identifying potential emergency situations, is provided in Section 6.2 of this guide.

By planning actions to address the most important risks and opportunities in its EMS, an organization will be taking steps to achieving its intended outcomes. Establishing an effective process (or processes) for identifying and prioritizing its environmental aspects, its compliance obligations and other organizational issues ensures consistent and reliable results. The organization should establish criteria for determining the most important ones to tackle. Documenting the methodology used is one way to ensure that the process (or processes) are carried out as planned. A process map, procedure or work instruction, and risk matrix (see Figure 5) are examples of ways to document the methods and criteria used. In addition to documenting the methodology, ISO 14001 also requires the results be documented.

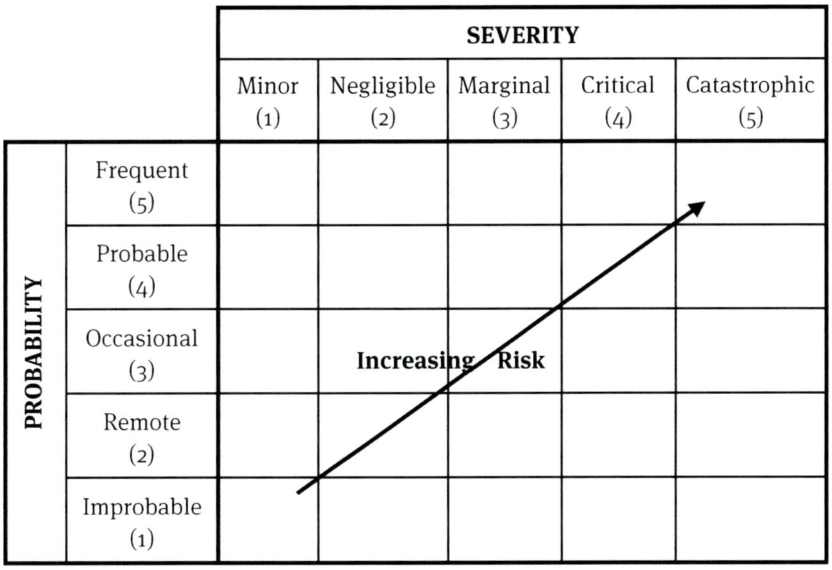

Figure 5 — **Example of a risk matrix**

There is no single approach or order to determining the risks and opportunities that need to be addressed. For ease in understanding and application, this guide provides guidance as follows

a) Determining the adverse and beneficial impacts on the environment caused by the organization's environmental aspects (Section 4.2).

b) Addressing compliance obligations (Section 4.3).

c) Identifying additional priority risks and opportunities the organization determines need to be addressed, caused by any of the other issues identified in its organizational context or are related to interested party needs and expectations that can have a detrimental or beneficial effect on achieving the EMS intended outcomes (Section 4.4).

NOTE This order differs from the order presented in ISO 14001:2015, in that organization context risks and opportunities are presented first (Clause 6.1.1), followed by environmental aspects (Clause 6.1.2) and then compliance obligations (Clause 6.1.3).

An organization can choose to follow this approach or choose another approach. For example, identifying risks and opportunities can be combined with the identification of environmental aspects and/or compliance obligations. Some organizations combine the identification of risks and opportunities with understanding organizational context and interested parties. Each organization can decide on an approach that makes sense for them; then streamline it and refine it over time.

4.2 Environmental aspects and environmental impacts (ISO 14001, Clause 6.1.2)

Environmental aspects are the elements of an organization's activities, products or services that can interact with and potentially affect the local or regional condition of the air, water, land, flora and fauna. Simply put, an organization's environmental aspects can cause environmental impacts. These impacts can be beneficial or adverse. Heating a facility (an activity) might cause carbon dioxide emissions from burning fossil fuels in a boiler (an environmental aspect), and can impact the environment by contributing to air pollution (an environmental impact). An organization can also contribute positive benefits to the environment. Use of natural resources (an environmental aspect) can be preserved (a beneficial environmental impact to wildlife habitat) by setting aside protected areas from future development (an activity).

The potential harmful effects an organization's significant environmental aspects have on the environment can be broadly thought of as "risks". Alternatively, the potential positive effects they have can be thought of as "opportunities". Historically ISO 14001 referred to these effects as adverse environmental impacts and beneficial environmental impacts, respectively. An organization needs to understand the way in which its activities, products and services can cause an adverse environmental impact, so that it can plan actions to minimize its risks. Similarly, it is necessary to understand if there are any beneficial environmental impacts it can have so it can pursue opportunities to improve the environment.

4.2.1 Getting started

4.2.1.1 Taking stock of your current state

Most organizations have some knowledge of their environmental aspects. Environmental laws and regulations address specific environmental aspects to protect against adverse environmental impacts. Facility-specific environmental permits issued by regulatory authorities often point to specific activities or equipment that need to be controlled to prevent unacceptable impacts. But not all environmental aspects are necessarily regulated in all countries (e.g. energy use, water consumption and material use). Information on pressing local or regional environmental conditions can be sought from industry groups or universities. Safety data sheets that accompany purchased chemicals also provide information on environmental aspects.

It is necessary to not only evaluate what the organization is currently doing and producing but also what it has done in the past or plans to do in the future. There may be environmental impacts associated with past practices that may need to be remediated. Looking proactively, the organization has the option to avoid generating environmental aspects when planning new products.

4.2.1.2 Overview

Identification of environmental aspects, and the associated impacts, applies to all activities, products and services within the EMS scope, including its facilities and infrastructure. It covers routine operations, as well as non-routine or abnormal circumstances such as start-ups, shut-downs and emergencies. It also includes any planned changes to its activities, products or services. The evaluation includes those performed both on site as well as those performed off site. Offsite examples are contracted services, maintenance activities performed at a customer location, travel services, and product demonstration activities by the sales force.

There are also environmental aspects associated with each stage of the product life cycle that need to be considered, i.e.

- acquisition of raw materials,
- design,
- production or manufacturing processes,
- transport of the product to market or delivery of service to the customer,
- use of the product or service,
- end-of-life treatment and final disposal.

In ISO 14001:2015, this is referred to as a "life cycle perspective". The intent is to look at the consecutive and interlinked stages of a product, often referred to as "cradle to grave", to identify environmental aspects that the organization can control or influence.

Some environmental aspects are under the direct control of the organization; others are not but in some cases can be influenced; still others cannot be controlled or influenced. An organization can control the amount of electricity it uses, and in some cases where it buys the electricity from, but it is unlikely that it can control or influence the way the electricity is generated by its local power company. A travel agency making reservations for its clients cannot control the emissions of the airplane, however it may influence the clients' mode of travel by providing information on the greenhouse gases generated for their flight. When an organization designs its products for direct sale to consumers, it can control the raw materials, substituting less hazardous materials, and influence the consumer on available recycling options or proper disposal methods. How the supply chain produces goods and services is not within the organization's control, but the choice of supplier is. An organization can select suppliers based on their environmental practices. They can stipulate environmental requirements for goods purchased in procurement documents. By leveraging the law of supply and demand, organizations can influence the supply chain by demanding environmentally sound products.

Organizations are expected to focus on identifying the environmental aspects it can control and those it decides it can influence. Each organization will need to determine for itself what to influence. Factors that play a part in those decisions include legal or contractual authority, its relative size and capability and its own environmental requirements embodied in its policies.

Identifying environmental aspects is an ongoing process. As the organization adapts its operating procedures, equipment, energy and raw materials purchases, or grows its products and services, the environmental aspects need to be re-assessed to ensure that it is focusing on the priority ones. The rate of change within the organization's activities, products and services will dictate how often the environmental aspects will be re-assessed. When changes occur often, a quarterly review, for example, may be prudent; when there is stability, an annual review may be sufficient.

NOTE The methodology and criteria for determining significant environmental aspects is required to be documented and kept up-to-date.

Records of its environmental aspects, associated environmental impacts and the significant environmental aspects are required to be retained.

Communicating significant environmental aspects

Communicate within the organization, in particular to
- those assigned responsibility for achieving a related objective;
- those whose work involves a significant environmental aspect.

Report environmental aspect information
- required by regulatory authorities;
- to relevant interested parties which the organization decides.

Communicate to others as prescribed in the organization's communication process.

4.2.1.3 Environmental aspects and impacts

Environmental aspects are the ways the organization interacts with the environment. The materials and natural resources used and the products or by-products from its operations create environmental aspects. An environmental impact is a change to environment, whether adverse or beneficial that results from an environmental aspect. Simply stated, the relationship between an aspect and impact is one of cause and effect; environmental aspects cause environmental impacts.

Table 4 provides a list of the most typical environmental aspects and some of the associated potential adverse and beneficial environmental impacts.

Table 4 — Environmental aspects and potential environmental impacts

Environmental aspect	"RISKS" Adverse environmental impact	"OPPORTUNITIES" Beneficial environmental impact
Air emissions	Air pollution	
Waste water discharges	Water pollution, degradation of aquatic habitats, degradation of drinking water supply	Source elimination, substitution, reduction or reuse
Spills/releases to land	Soil pollution, groundwater pollution	Remediation of pollution, to repurpose the land
Chemical/hazardous material use	Soil pollution, water pollution, air pollution, bioaccumulation of toxic substances in fauna; health effects to employees and local residents	Replacement of pesticides with biological controls and beneficial insects
Raw material use	—	Conservation of raw materials; use of recycled content materials
Natural resource use (water, land, trees, minerals)	Depletion of natural resources, degradation of biodiversity and habitats, land contamination	Conservation; preservation of natural habitats; waste water reuse
Electricity consumption	Air pollution, global warming, land contamination	Use of renewable energy sources (wind, solar, etc.)
Fuel use	Depletion of non-renewable fossil fuels	Conversion to alternative fuel sources (ie. fuel cells, methanol, biomass)
Waste generation	Soil pollution, groundwater pollution	Waste minimization; waste-to-energy recovery technology
Reuseable/recyclable waste products and by-products	—	Re-purposing of by-products and wastes; Reusable packaging totes and pallets; reuse/salvage building materials
Noise, odour, dust, traffic	Discomfort or inconvenience to local residents	

An appraisal of environmental aspects and impacts may require technical reviews and informed judgement to achieve credible results. Resources may be readily available within the organization from safety data sheets, monitoring data, regulatory permits or licenses, emergency situations that occurred in the past, product specifications and bill of materials. Additional technical resources may be needed and are available through local universities, industry groups and regulatory agencies.

4.2.1.4 Significance

Since an organization can have many environmental aspects that can impact the environment in a variety of ways, it is necessary to determine the ones that need to be addressed. In ISO 14001 these are called "significant environmental aspects". What is significant to one organization, however, may not be significant to another. The concept of significance is a relative one, and will depend primarily on the severity and frequency of its environmental impacts, and may also depend on the views of external interested parties, the organization's compliance obligations or its resource capabilities. Moreover, the environmental aspects that an organization deems significant may change over time. As process improvements are implemented and environmental aspects are controlled, additional aspects with less severe or less likely environmental impacts can be identified as significant and incorporated into the EMS. Therefore, it is important that a consistent and reliable methodology for determining significant environmental aspects be established.

The methodology can be quite simple and intuitive, such as asking a series of questions. For example, "Will it pollute a nearby waterway?" or, "Is it a regulated environmental aspect?" or, "Is there a local or regional problem concerning this aspect?". If the answer is "yes", then it can be deemed significant; if the answer is "no", then it is not.

Conversely, a more detailed methodology can be applied using information, such as the scale, severity and duration of the environmental impact, or type, size and frequency of the environmental aspect. A qualitative ranking, such as "high-medium-low" or a quantitative scoring (i.e. 1-10) can be applied to each criterion, then rank ordered. Using a pareto analysis or establishing a threshold value helps to determine the most important environmental aspects to address first, and which can be tackled later, over time. Although not required, organizations may use formal risk assessment methodologies if they have the technical competence to apply them correctly.

Environmental criteria is the primary criteria for determining significance, but there may be other non-environmental risks and opportunities to the organization that can be considered. For example, an air emission that was not deemed significant by environmental criteria alone, may be significant if neighbours have filed complaints about the smoke or odours. Similarly, if the air emission currently meets applicable legal requirements, but increased production is anticipated causing additional emissions, then it may be deemed significant in order to control the emission level within existing limits. Consideration of these other issues can be incorporated in the significance methodology, or can be considered separately when determining other risks and opportunities to the organization, as described in Section 4.4.

The process of determining significant environmental aspects is very important as it forms the technical basis for the EMS. Those determined to be "significant" warrant management within the EMS, and are subject to its requirements. That is,

- Applying the policy commitments to protect the environment, comply with obligations and continually improve.
- Planning actions to address the environmental impact, including establishing environmental objectives.
- Ensuring competency, training and awareness of environmental requirements.
- Allocating responsibility, authority and resources for their control.
- Monitoring, measuring and analysing performance.
- Reviewing changes, and evaluating whether the actions planned will continue to achieve the desired results.

Key linkages

- Organizational context
- Environmental policy
- Planning action
- Environmental objectives
- Competence and awareness
- Operational control
- Monitoring, measurement, analysis and evaluation
- Emergency preparedness and response
- Management review

4.2.2 Helpful hints

It can be helpful to identify the range of the organization's activities, products and services within the scope of its EMS. An organization chart showing each department or unit, or a map of the facility and surrounding area, are good tools to refer to when first identifying activities. Then, identify the services used or performed related to each activity. Lastly, identify the products produced. This yields a pretty comprehensive set of activities, products and services to begin identifying environmental aspects. Additions can be made later as the organisation becomes more familiar with the process.

It is not necessary to evaluate the environmental aspects of each product, component or raw material individually. An organization can group them into product groups when they have common characteristics, such as organizational units, geographical locations, operations workflow, materials or energy use, or environmental media affected. To be useful, however, the size of a grouping should be large enough for meaningful examination, yet small enough to be clearly understood.

Input from employees across all functions (e.g. sales, manufacturing, operations, finance, maintenance) can help provide a complete picture of its activities and the associated environmental aspects and impacts.

There is no single way to identify significant environmental aspects. Each organization needs to select a method that is appropriate to the scope, nature and scale of its EMS and that suits its needs. Organizations needs to balance the level of detail needed, the complexity of the methodology, and the time and cost necessary to make a credible and reliable assessment. It is important, however, that a systematic approach is used to make sure all activities, products and services within the EMS scope are covered, and the criteria to determine significance is applied in a consistent way.

More detailed information on determining significant environmental aspects can be found in ISO 14004, Clause 6.1.2.

4.2.3 Check conformity

Requirement	Y	N	Follow-up actions
Identify environmental aspects of its activities, products and services:			
• include abnormal conditions and emergency situations,			
• consider each stage of the product life cycle,			
• include planned changes.			
Determine if the environmental aspect can be controlled.			
Determine if the environmental aspect can be influenced.			
Document the environmental aspects.			
Determine associated environmental impacts.			
Document the environmental impacts.			
Establish criteria for determining significance.			
Determine its significant environmental aspects.			
Document its significant environmental aspects.			
Document the methodology and process used.			

Requirement	Y	N	Follow-up actions
Communicate its significant environmental aspects to relevant levels and functions within the organization.			
Periodically review and reassess environmental aspects:			
• including those resulting from new or modified activities products or services.			
Update documentation as needed.			

4.3 Compliance obligations (ISO 14001, Clause 6.1.3)

In order to meet both the environmental policy commitment and the EMS intended outcome to fulfil compliance obligations, an organization needs to have detailed knowledge of their obligations and how they affect its operations. These include both environmental-related legal requirements and other requirements that the organization has to or chooses to comply with. Since changes occur over time, to both operations and the requirements themselves, the organization needs a method for keeping their requirements updated.

Further, the EMS must be designed to fully address these compliance obligations. Controls need to be in place to ensure the organization operates in compliance with the requirements and actions must be taken when compliance obligations are not met. This is important for ensuring violations and fines for noncompliance do not occur.

4.3.1 Getting started

4.3.1.1 Taking stock of your current state

Permits and licenses to operate are a good starting point for identifying legal requirements. Compliance audit reports performed by regulators or consultants also are good sources of information.

Determine if the organization has any non-regulatory requirements. There may be corporate environmental policies or customer requirements that must be satisfied. Customers might have requested its suppliers to report environmental information to a third party. The organization may belong to industry or trade groups that have environmental provisions required of its members. Consider any voluntary agreements made with community groups to address particular environmental concerns. Additionally, if the organization has agreed to participate in voluntary initiatives or reporting schemes that are related to environmental management, these will also need to be included. Don't overlook the obvious voluntary initiative to implement ISO 14001.

The process of identifying and maintaining knowledge of environmental requirements is not unlike the process used for other types of requirements, like tax, import/export, occupational safety and employment laws, or customer contract requirements. To the extent that those processes work well, they can serve as the basis for implementing and tracking compliance obligations.

4.3.1.2 Overview

Compliance obligations are legal requirements that an organization has to comply with and other requirements that an organization has to or chooses to comply with. In terms of ISO 14001, it includes only those requirements that are related to its EMS, and more specifically, to the organization's environmental aspects. They arise from mandatory requirements, such as applicable environmental laws and regulations, or voluntary commitments made by the organization, such as the Business Charter of Sustainable Development. Table 5 gives examples of each.

Table 5 — Examples of compliance obligations

Legal requirements	Other requirements
Requirements from governmental entities or other relevant authorities	Agreements with community groups or non-governmental organizations
International, national and local laws and regulations	Agreements with public authorities or customers
Requirements specified in permits, licenses or other forms of authorization	Organizational requirements
Orders, rules or guidance from regulatory agencies	Voluntary principles or codes of practice

Legal requirements	Other requirements
Judgements of courts or administrative tribunals	Voluntary labelling or environmental commitments
Legal requirements imposed by government authorities where products are sold	Contractual arrangements
	Relevant organizational or industry standards

Based on the decision made in Section 2.2, the organization has a high level understanding of the interested party requirements it has to or chooses to comply with. Next the organization drills down on the specific provisions to a level of detail sufficient to determine how they apply. Some provisions will relate to specific operations, such as stipulating hours of operations, discharge limits or required monitoring; while others are general provisions related to the organization as a whole, such as reporting requirements, necessary training or recordkeeping.

Knowing what the requirements are and how they apply enable the organization to implement the necessary steps to ensure their operations are conducted in a manner consistent with the requirements. Although all elements of the EMS support achievement of regulatory compliance (see Section 4.3.1.3), the key elements an organization needs to establish are operational control of processes and equipment (see Section 6.1), performance monitoring (see Section 7.1), compliance evaluation (see Section 7.2), competency (see Section 5.2), awareness (see Section 5.3) and documented information (see Section 5.5).

NOTE A list of the compliance obligations is required to be documented and kept up-to-date.

The organization should also document the processes used to ensure consistent implementation.

As with many of the other elements of an EMS, maintaining compliance obligations is an ongoing activity. Requirements change over time, as do operations. It is important to be sure the organization has access to current information on how compliance obligations apply to their operations. An up-to-date list of the compliance obligations is necessary to be sure changes can monitored.

Including the origin, the applicable environmental aspect and the operation it applies to may also be helpful.

4.3.1.3 Managing regulatory compliance using an EMS

For some small businesses, the process of identifying applicable compliance obligations will be relatively simple. Depending on the country in which they operate, their air emissions, water discharges and wastes generated may fall below regulatory thresholds. They may be subject to general provisions in environmental law, however, such as those related to disposal of wastes. They may be handled relatively easily within their operations. For others, the applicable laws and permit conditions are numerous and complex. They struggle to understand the provisions and how they apply, not to mention sustaining operations in compliance with their legal requirements. For these organizations, an EMS can be used to focus on attaining and sustaining compliance.

Managing regulatory compliance in either case can be done systematically and effectively by applying the EMS elements, but most importantly those described below.

- Determine the legal requirements related to its environmental aspects that the organization has to comply with (ISO 14001, Clause 4.2). Determine their applicability and maintain information on existing, new and changing requirements. (ISO 14001, Clause 6.1.3).
- Determine risks and opportunities associated with legal requirements that need to be addressed, and plan actions to assure compliance, integrating the actions into its business processes (ISO 14001, Clauses 6.1.1 and 6.1.4).
- In cases where complying with a legal requirement will require time, establish an objective and action plan to address the compliance issue (ISO 14001, Clause 6.2).
- Ensure persons doing work under the organization's control that can affect its compliance status are competent and have completed any regulatory required training. Ensure they are aware of the EMS compliance related requirements and their role in meeting those requirements (ISO 14001, Clauses 7.2 and 7.3).

- Communicate with the authorities and file regulatory reports on time (ISO 14001, Clause 7.4.3).
- Create and control documented information, including records, necessary to demonstrate compliance with legal requirements. Identify and maintain control of external documents generated by regulatory agencies that are needed to plan and operate the organization's EMS (ISO 14001, Clause 7.5).
- Establish, implement, control and maintain the processes necessary to meet compliance-related EMS requirements and environmental objectives. (ISO 14001, Clause 8.1).
- Monitor or measure indicators of legal compliance, then analyse and evaluate the organization's performance (ISO 14001, Clause 9.1). Periodically evaluate compliance with legal requirements. (ISO 14001, Clause 9.1.2).
- Take prompt action to correct non-compliant situations and compliance-related EMS nonconformities, and implement corrective action(s) to address the cause and prevent reoccurrence (ISO 14001, Clause 10.2).
- Take action to improve the organization's compliance status, including improving the compliance-related EMS processes (ISO 14001, Clauses 10.1 and 10.3).

Taken together, these provisions enable an organization implementing ISO 14001 to systematically identify and manage its legal requirements consistent with top management's commitment to fulfil its compliance obligations.

4.3.2 Helpful hints

The Internet is a useful tool for obtaining information on environmental laws and regulations. Access to the regulations generated by regulatory bodies is often available online.

Building knowledge of compliance obligations internally is best and can be done by attending courses and seminars. Maintaining knowledge of changing regulations can be accomplished through online commercial services, regulatory agencies, trade groups and regulatory newsletters. Alternatively, external resources such as consultants, attorneys, industry groups and small business organizations can help with this. Often regulatory bodies offer assistance to small businesses.

Augmenting the list of applicable compliance obligations can be valuable. Provide a brief summary of requirement, the source, a description of the applicability to the organization and the function/unit/role responsible for managing it. Include a reference to where more detailed information can be found. The list can be used to track requirements and communicate responsibilities. It can also be handy when evaluating compliance (Section 7.2).

Often organizations focus only on the technical requirements that apply to operations or facility equipment. They overlook the administrative requirements. Filing regulatory reports on time can result in violations, and missing deadlines on voluntary reporting can discredit the organization. A tickler system or calendar reminders can be used to track administrative requirements.

4.3.3 Check conformity

Requirement	Y	N	Follow-up actions
Determine the mandatory compliance obligations related to its environmental aspects.			
Determine the voluntary compliance obligations related to its environmental aspects.			
Determine the applicability of compliance obligations to the various operations and functions within the organization.			
Document the compliance obligations.			
Establish a mechanism (or mechanisms) to keep informed of the requirements associated with the organization's compliance obligations.			
Periodically review applicability of new or changed requirements, and make changes as required to be in compliance.			
Update documented information as needed.			

4.4 Other risks and opportunities (ISO 14001, Clause 6.1.1)

Although an EMS is primarily focused on the way an organization affects the environment, with emphasis on the risks and opportunities associated with its significant environmental aspects, there could be other issues associated with the organizational context that can either help or hinder the organization from achieving the intended outcomes of its EMS. The local and regional environmental conditions can affect its operations. Government entities may curtail business operations when ozone or particulate levels exceed certain levels. The manner in which an organization attends to its compliance obligations affects its compliance status and in turn, its reputation and brand. Other considerations, both within and outside the organization's control such as staff turnover rate and economic factors, can affect the EMS.

In many cases, the adverse effects are often the first to be unveiled. Being aware that risks exist enables an organization to plan ways to overcome the adversity created, and even in some cases, turn the situation around to create an opportunity with an overall beneficial effect.

Identifying the important risks and opportunities associated with these other issues, the ones that *need to be addressed*, enables an organization to not only achieve its policy commitments and environmental objectives, but also protect its business and reputation so it can continue to operate and grow.

4.4.1 Getting started

4.4.1.1 Taking stock of your current state

Are there any risk assessment tools or methods currently in use in the organization? Adapting a familiar methodology to environmental issues may be easier to implement than taking a new approach.

A conversation is necessary among top management on their willingness or unwillingness to address potential effects when the consequences and likelihood it will happen is uncertain. This will aid in assessing the organization's "risk appetite" and help determine which risks and opportunities need to be addressed.

4.4.1.2 Overview

Since the organization has already addressed adverse and beneficial environmental impacts associated with its environmental aspects to determine the ones that are significant, it has already addressed the most important set of risks and opportunities to the EMS. However, there may be others. The two sets of inputs to consider when determining if there are other priority risks and opportunities that need to be addressed are: i) the relevant issues identified from understanding the organizational context, and ii) the selection of compliance obligations. The organization should focus on those that

- **a)** facilitate achievement of the EMS intended outcomes;
- **b)** prevent, or reduce, undesired effects;
- **c)** achieve continual improvement.

Table 6 — Two-step process for determining which risks and opportunities need to be addressed

Step 1. What can happen? Revisit the relevant issues that were identified when understanding the organization's context. Start by thinking of what can happen to the EMS or its intended outcomes. How will that effect the organization as a result? The more people involved in the brainstorming the better. "What if" exercises can help to capture the obvious adverse and beneficial effects and trigger further scenarios.
Step 2. What's the effect? Is it positive or negative? How much so? How likely is it? If the right people are involved, a qualitative estimate of the potential impact and how likely it is to occur is all that is needed. Simply express it on a pre-defined scale of low-medium-high. Plotting these on a matrix can visually show which are high impact and high likelihood, and therefore need to be addressed.

When starting out, there may be many issues facing the organization, so following a simple process will be more than adequate, like brainstorming and asking "what if" questions, as described in Table 6. If the tool "Strengths, Weaknesses, Opportunities and Threats analyses" (SWOT) was used to determine the organizational context, continue to use it to prioritize the issues identified. ISO 14001 does not require a quantitative risk assessment when a qualitative assessment can suffice. As the programme matures and evolves, as the obvious risks and opportunities are addressed over time, more sophisticated methodologies can be pursued, such as "Political, Economic, Social, Technological, Legal, Environmental analysis" (PESTLE), a probability/severity risk matrix, or "Failure, Mode and Effects Analysis" (FMEA). Considerations when choosing any tool include i) the competency needed to use the tool correctly, ii) time and costs, iii) number and complexity of issues to assess. The organization should select a method that suits their needs.

Keep in mind that documentation of every issue and every interested party requirement that is evaluated for a potential adverse or beneficial effect is not required. ISO 14001 does, however, require the organization to maintain documented information on the risks and opportunities it determines needs to be addressed. Therefore the outputs of the analysis, at a minimum, needs to be documented.

NOTE Records of risks and opportunities that need to be addressed are required to be retained.

For consistency purposes, the process for determining risks and opportunities that need to be addressed should be documented and kept up-to-date.

There are many different approaches an organization may take when assessing these other issues and compliance obligations for risks and opportunities. It is not necessary to conduct separate analyses; the analyses can be combined in whole or in part with the analyses of significant environmental aspects or the analysis of compliance obligations or the analysis of the organization's context. For example, the process can entail one of the following approaches, or others that suits the organization's needs, as described in ISO 14004, Clause 6.1.1:

- determining environmental aspects, compliance obligations and other issues and requirements separately, then determining the risks and opportunities that need to be addressed for each of these; or
- integrating the determination of risks and opportunities that need to be addressed into its determination of significant environmental aspects, its analysis of compliance obligations, its identification of issues associated with organizational context, and its determination of interested party needs and expectations;

- determining risks and opportunities in combination with the process the organization uses to gain an understanding of its context;
- follow an alternative approach where two or more of the sources of risks and opportunities are considered in combination to determine which need to be addressed.

The organization identifies and prioritizes its risks and opportunities, then selects the ones that it determines need to be addressed. The organization then plans the appropriate action within its EMS to mitigate any adverse effects and pursue the positive effects (see Section 4.4). However, each organization is unique, therefore it needs to determine for itself if there are any other risks and opportunities it needs to address.

For illustrative purposes, the following broad categories will be explored, providing examples of how they can pose risks and opportunities to the EMS and achievement of its intended outcomes:

- environmental conditions,
- compliance obligations, and
- other risks and opportunities.

Key linkages

- Organizational context
- Interested party requirements
- Environmental aspects
- Compliance obligations
- Planning action
- Emergency preparedness and response

4.4.1.3 Risks and opportunities associated with environmental conditions

Local and regional environmental conditions can affect an organization's operations, and in some cases this effect can be detrimental. For example, eroded marshlands can cause flooding during heavy rainfall to businesses located on the seashore, potentially shutting down operations and causing damage to the facility infrastructure. Discussion of marshland mitigation strategies with management can be important when they are making decisions whether to rebuild or expand its facilities.

High levels of smog can result in a temporary shutdown of business by regulatory agencies during peak NO_X and O_3 concentrations. If unexpected and unplanned for, businesses could lose revenue while they are shut down. However, regulators may offer incentives to businesses who voluntarily offer to reduce their energy load during these peak periods. An organization can turn a potential negative effect into a positive one if it can shift its production to off-peak hours.

Organizations operating in water scarce regions of the world will be challenged to reimagine and redesign processes that are water intensive, by considering alternative cooling liquids or maximizing closed loop systems. Innovative processes, approaches and technologies are born when the organization turns adversity into opportunity.

4.4.1.4 Risks and opportunities associated with compliance obligations

There are adverse effects associated with non-compliance with regulatory requirements. For example, there is possible damage to the environment for which the organization will be required to make compensation, either in terms of remediation or payment of fines. Further, these can end up in the newspaper and negatively impact the organization's public image. Regulatory fines and violations associated with not complying with laws or permit conditions, can erode confidence in the organization, affecting its brand or potentially losing its license to operate.

The selection of relevant interested party needs and expectations has the potential to effect the organization. Those selected will be addressed in the EMS, which will be viewed positively by the respective stakeholder. On the other hand, interested parties whose requirements are not being addressed may not view the organization well. An unresolved need and expectation not being addressed by the organization poses a potential for an adverse effect.

On the other hand, the organization can garner beneficial effects in dealing with its compliance obligations. Organizations that detect new regulations on the horizon can take the opportunity to modify their activities in advance of it taking effect, perhaps avoiding applicability of the regulation and saving the organization ongoing compliance costs. An organization may take a proactive stance, going above and beyond what is required by law by controlling its discharges at levels lower than its permit limits, thereby improving its environmental performance and perhaps being recognized as an environmental leader in the community with an award.

4.4.1.5 Other risks and opportunities

Emergency situations are typically viewed as low probability events and therefore not addressed. However when they happen, the effects on the organization can be devastating. Uncontrolled releases of hazardous substances can erode confidence in the organization, potentially losing its license to operate.

Economic conditions can impact access to financing or affect an organization's financial position. In some cases, access to capital is restricted, which can delay progress on an environmental objective requiring installation of new environmental technologies. On the other hand, when interest rates are low, it may be easier and cheaper to acquire capital.

Organizational changes can affect the EMS. For example, depending on how an organization approaches the following, the EMS intended outcomes can be positively or negatively affected:

- changes in production causing additional shifts or a large increase in contract labour,
- changes in management or organizational structure, and
- acquisitions or divestitures.

An organization may find that some of its environmental aspects do not meet the threshold for significance when considering environmental criteria alone, but when interested party concerns are added to the evaluation, these aspects may need to be addressed. For example, noise levels may not reach the threshold of a significant environmental aspect but when taking into consideration complaints from neighbouring community members, the risks and opportunities may warrant being addressed.

4.4.2 Helpful hints

Keep it simple. Often an intuitive approach yields the same general results as applying a risk assessment methodology.

The EMS should focus on significant environmental aspects and compliance obligations first and foremost; it's possible that an organization does not have any other risks and opportunities that need to be addressed.

4.4.3 Check conformity

Requirement	Y	N	Follow-up actions
The organization has a process for determining risks and opportunities to the organization related to other organizational context issues and/or interested party requirements.			
The process considers:			
• Environmental conditions that can affect it.			
• Compliance obligations, and other interested party needs and expectations.			
• Other relevant issues associated with its context.			
The process prioritizes risks and opportunities based on:			
• Ability to achieve the intended outcomes.			
• Preventing or reducing undesired effects.			
• Achieving continual improvement.			
Documented information of the following is maintained:			
• Risks and opportunities to the organization that need to be addressed.			
• Process used.			

4.5 Planning action (ISO 14001, Clause 6.1.4)

At this point an organization has determined its significant environmental aspects, the applicability of its compliance obligations and other priority risks and opportunities that need to be addressed. Now, it is ready to plan how it will address these matters in their EMS. The actions planned should reflect the decisions made on how best to treat the matter, taking into account business considerations.

The actions are integrated into the detailed planning of its EMS processes as appropriate, or into other business processes. The organization also determines how it will evaluate the effectiveness of the planned action so that follow-up can occur.

4.5.1 Getting started

4.5.1.1 Taking stock of your current state

Has the organization already implemented actions to manage some environmental aspects, compliance obligations, or risks and opportunities? If needed, build on what is already being done.

Can the organization's business planning processes be adapted to accommodate the actions that need to be planned for the EMS?

4.5.1.2 Overview

When planning the actions needed to address significant environmental aspects, compliance obligations and other risks and opportunities, the organization needs to factor in business considerations, such as:

- technological options available, and their feasibility;
- costs and overall availability and constraints of financial resources;
- organizational strategies, priorities and plans;
- production and other operations/functions requirements.

The plan of action that an organization takes can entail a single action or multiple actions. It may be sufficient to leverage the beneficial impact of using paper with 30 % recycled content by instituting an administrative procurement control. In another case, an environmental objective and detailed plan may be necessary to upgrade equipment and operations to address new regulations.

The organization should consider how the elements of the EMS can be used to address the needed action, for example:

- setting environmental objectives and actions plans, to mitigate adverse effects or leverage beneficial opportunities
- implementing hierarchy of operational controls (see Figure 6), by eliminating or mitigating the source
- enhancing emergency preparedness and response processes so it is prepared to mitigate the adverse environmental impact if it should occur
- providing written instructions to consumers on reuse and recycling methods, to leverage beneficial environmental impacts
- monitoring and measurement, taking no proactive measures other than watching how it evolves.

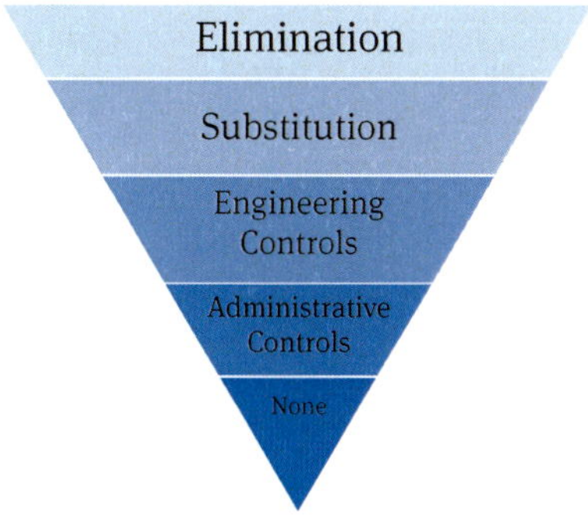

Figure 6 — Hierarchy of control

At this point, the organization need not perform detailed planning of these actions. It is sufficient to just identify the course of action. For example, the organization can choose to set an environmental objective. The detailed planning of the tactical actions needed for that objective will come later. Examples for illustration purposes are given in Table 7.

NOTE For consistency purposes, the process for planning action should be documented and kept up-to-date.

The actions planned should be commensurate with the organizational effect or environmental impacts. The more severe the effect or impact, the more comprehensive the actions planned should be.

Key linkages

- Environmental aspects
- Compliance obligations
- Risks and opportunities
- Environmental objectives
- Resources
- Competence
- Awareness
- Communication
- Operational control
- Emergency preparedness and response
- Monitoring, measurement, analysis and evaluation
- Evaluation of compliance
- Internal audit
- Management review

4.5.1.3 Evaluating effectiveness

In simple terms, to check the effectiveness of actions requires the organization to ask, "Does it work?". There are various methods an organization can check the effectiveness of actions, including:

- monitoring status,
- analysis of results using performance indicators,
- audits,
- management reviews, and
- nonconformity and corrective action.

4.5.2 Helpful hints

Coordinate schedules of EMS planning with the business planning cycle.

It is important to remember that the requirement to plan action and evaluate effectiveness is not intended to duplicate the planning or evaluation that is done in interrelated clauses of the EMS, such as the clause on monitoring, measurement, analysis and evaluation. Instead, use those processes when addressing risks and opportunities and evaluating the effectiveness of the planned action.

4.5.3 Check conformity

Requirement	Y	N	Follow-up actions
The organization has planned to take action to address its significant environmental aspects.			
The organization has planned to take action to address its compliance obligations.			
The organization has planned to take action to address its other priority risks and opportunities to the organization and the EMS.			
The organization considers technological options and its financial, operational and business requirements when planning the actions.			
The organization has planned how to integrate these actions into the EMS processes, or other business processes.			
The organization has planned how to evaluate effectiveness of these actions.			

Table 7 — Illustrative examples of actions to address risks and opportunities (excerpts from ISO 14004:2015)

Activity/product/ service	Aspects	Risks and opportunities	Planning to take action
Consumer use of air conditioning unit	Use of electricity	**Adverse effects** • Depletion of non-renewable natural resources	Benchmark performance against other competitors. Invest in more research and development in relation to energy efficiency.
	Use of refrigerants	**Adverse effects** • Ozone depletion when air condition system leaks **Beneficial effects** • New service offering qualified engineers	Establish partnership with research institute on alternatives to refrigerants.
Service: Maintenance and repair services			
Chemical handling and use	Uncontrolled release during fire or explosion (emergency condition)	**Adverse effects** • Clean-up costs • Fines • Detrimental publicity	Establish environmental objective to replace chemical used with less toxic alternatives.
Subcontracted air conditioner repair	Release of ozone depleting substances (i.e. refrigerant) (abnormal condition)	**Adverse effects** • Fines • Detrimental publicity	Retender contract to ensure improved maintenance.

Activity/product/ service	Aspects	Risks and opportunities	Planning to take action
Service: Office support services			
Two-sided copying	Natural resource consumption	**Beneficial effects** • Conservation of natural resources • Reduced costs	Develop marketing literature to promote environmental and cost benefits to potential customers.
Recycling waste paper	Solid waste generation	**Beneficial effects** • Reduction in waste to landfills • Reduced costs • Positive publicity	
Service: Transportation and distribution of products and services			
Routine fleet maintenance	NOX emissions Disposal of oily wastes	**Adverse effects** • Air and soil pollution **Beneficial effects** • Recycle oily wastes • Reduced operating costs	• Communicate environmental benefits to maintenance workers. • Develop operational control process to manage waste. • Consider switching to vehicles powered by electricity during re-capitalization.

Activity/product/service	Aspects	Risks and opportunities	Planning to take action
Fleet operation	Fuel consumption	**Adverse effects** • Fossil fuel availability • Higher fuel costs **Beneficial effects** • Use alternative fuel (CNG/LNG) • Reduce fuel costs	Establish environmental objective to reduce fuel consumption.
	Emission of oxides of nitrogen	**Adverse effects** • Stricter fuel emission standards	Research methods to reduce emissions.
	Generation of noise	**Adverse effects** • Discontented neighbours • Detrimental to image of organization	Meet with community members. Limit operating hours.
Compliance obligations: Regulatory reporting			
Regulator requests for information		**Adverse effects** Failure to respond, delayed response or inaccurate response can lead to greater scrutiny from regulatory agency **Beneficial effects** Timely, pro-active and transparent communication can strengthen the organization's relationship with regulators	Develop more effective communication processes for receiving and responding to communication from regulatory officials, including reporting schedules. Apply the internal audit programme to make recommendations to improve timeliness and transparency of communication, and take action for continual improvement of communication processes if needed.

Activity/product/ service	Aspects	Risks and opportunities	Planning to take action
Other issues and requirements: Environmental conditions affecting the organization			
Water scarcity (food and beverage sector organizations)		**Adverse effects** Production constraints due to limited availability of water **Beneficial effects** Increased efficiency realized through process optimization	Apply engineering controls to reduce water loss in the production process. Establish performance indicators and monitor/measure water use per unit of production.

4.6 Setting and achieving environmental objectives (ISO 14001, Clause 6.2)

Driving improvement starts with establishing a goal, developing an actionable plan, and then garnering support within an organization to achieve the goal. Goals can be broad based and apply across the whole organization, others can be focused on a particular department or process. Where appropriate, environmental goals should be aligned with and support the overall business goals. Developing environmental goals in balance with other business priorities facilitates allocation of resources and support for the environmental initiatives.

For an EMS, goals are called environmental objectives, and flow from the environmental policy. Selection of environmental objectives, therefore, focus on protecting the environment, satisfying compliance obligations or improving the EMS. Depending on its priorities in these areas, the organization sets the performance level it wants to achieve. Then it can plan the actions needed to achieve the results.

4.6.1 Getting started

4.6.1.1 Taking stock of your current state

Review the overall business goals. For instance, an organization that wants to reduce costs can benefit from environmental initiatives that reduce wastes, limit energy use or streamline bureaucratic administrative activities.

Understand the organization's strategic planning and budgeting cycle so that environmental goals can be approved and monies allocated at the same time business priorities are set.

Involve those who will be responsible for implementing the initiatives, so that action plans are comprehensive and have realistic timetables.

Key linkages

- Leadership
- Environmental policy
- Environmental aspects
- Compliance obligations
- Risks and opportunities
- Planning action
- Operational control
- Management review

4.6.1.2 Setting environmental objectives

Environmental objectives are typically established to make an improvement in an area previously identified as a priority. The evaluation of significant environmental aspects identified priority areas where efforts to protect the environment can be taken. Likewise, the detailed assessment of compliance obligations may have identified new regulations or areas that the organization needs to upgrade its operations to fulfil its obligations. Risks and opportunities the organization decides to address, particularly those that will take time to accomplish and require coordination among several functions, are good candidates.

These are the first places to look when setting environmental objectives, but not the only ones. The results from audits, management review and evaluation of compliance may be sources of areas needing improvement. In some cases, the organization may want to maintain a certain level of performance, so objectives don't always have to be set to improve.

When selecting environmental objectives, the organization should consider the skill level available, any financial or other resources required, and its other business objectives. Some discussion with interested parties may assist in making appropriate choices.

The organization will need to monitor its progress toward meeting its objectives. Performance indicators can be used to evaluate progress toward the targeted performance level. Indicators are objective and verifiable, and may take the form of thresholds for performance or activity ratios, for example:

- 50 % reduction in waste to landfill in the next two years.
- Achieving and sustaining a 96 % completion rate for required environmental training.
- Reducing GHG emission to 4 kilogrammes or less per tonne of product in three years.
- Eliminating substances of very high concern in products and processes (aka zero) in 5 years.
- Achieve and maintain fleet energy consumption rates at 14 kilometres travelled per litre fuel.
- Operating at levels 20 % lower than permit limits for pH and total suspended solids in waste water discharges.
- Increase ratio of green space to building footprint (m^2) in new construction by 20 %.

Objectives should be measurable, if practicable. Alternative approaches may be necessary to measure progress, for example:

It may not be practicable for an organization to measure GHG but can be estimated using energy consumption rates and applicable emission factors.

It may not be practicable to measure hazardous air pollutants directly, but can be estimated using chemical usage and hazardous material properties.

Although the initial phases of improvement projects may not yield measurable environmental performance improvement, progress can be measured by milestone achievement, such as 25 %, 50 %, 75 %, 100 % complete.

NOTE The environmental objectives are required to be documented and kept up-to-date. Records of monitoring progress on the objectives are required to be retained.

4.6.1.3 Action planning

Once the environmental objectives have been set, the organization can identify the actions necessary, assign responsibilities and establish milestones for achievement. Action planning is a dynamic activity and subject to amendment as objectives are revised or met.

Action planning can be relatively simple, identifying "who" needs to do "what" and "when" it needs to be done. If financial, equipment or external resources are needed to complete the tasks, they also need to be identified. An organization might find it helpful to use a simple template as shown in Table 8 to monitor progress on its environmental objectives.

Table 8 — Action planning template

Objective #1:					
Action	Responsibility	Due date	Resources needed	Progress	Follow-up needed?

More complex objectives, especially those that require a multi-year plan, involve capital investment or incorporate new technologies, will pass through several phases, e.g. a study of new equipment to be procured; evaluation of alternatives, costs and benefits; loan agreements with banks; coordination with original equipment manufacturers for delivery, commissioning and operation of new equipment; facility modifications and employee training. In these cases, it can be difficult to anticipate all of the factors that will affect decisions along the way. A high level plan of action can be developed covering the major milestones with detailed planning on the initial phase. As the organization advances toward completion of one milestone and implementation of the next approaches, detailed planning can be added for the next milestone. So on and so forth, creating a living plan. Each portion of the action plan needs to cover who, what, when and resources needed. Project management software can help track more complex action plans.

As progress is made, it will be important to sustain the gains achieved by institutionalizing the modified work processes into standard practices. As part of managing change, re-evaluate objectives and actions plans periodically particularly when changes are made to products, services and facilities, when new compliance obligations are introduced or changed, or when substantial progress has, or has not, been made on environmental objectives.

> Communicate the environmental objectives throughout the organization. Keep them updated on progress.
>
> Consider communicating objectives set and progress made in annual reports.

4.6.2 Helpful hints

Objectives which are too difficult to achieve can be discouraging and can weaken the perceived value of smaller milestones achieved along the way. Equally, if too long a timeframe is set for achievement, the objective can have a diminished significance and may even be forgotten.

Consultation with interested parties such as banks, insurance companies, clients, regulators and service engineers can help in setting priorities. Such consultations may also help with:

- Financial support in the form of concessions or grants
- Insurance premium reductions for environmental improvements
- Early indications of changing client requirements
- Identifying alterations to existing legislation and the impact of new regulations.

Enlist help from onsite contractors and suppliers to provide ideas on alternative materials or better ways of working that will support achievement of environmental objectives.

A test for a good action plan is to estimate the sum total reduction from each of the actions to verify the plan will yield the desired improvement.

4.6.3 Check conformity

Requirement	Y	N	Follow-up actions
Environmental objectives address priority areas associated with • Significant environmental aspects • Compliance obligations or • Other risks and opportunities			
Environmental objectives support the policy commitments to protect the environment, fulfil compliance obligations or continually improve the EMS.			
Measurable environmental objectives are established, documented and communicated.			
A plan of action is developed to achieve the environmental objectives describing • what actions/tasks need to be done • who is responsible • when the actions/tasks are due • resources required			
Where it makes sense, the actions to achieve environmental objectives are integrated into existing business processes.			
Progress is monitored and evaluated using indicators.			

5 Establishing support elements in an EMS

Establishing support elements is important to a successfully functioning EMS. These elements are part of the "Do" phase of the Plan-Do-Check-Act model. ISO 14001 identifies five clauses necessary for the EMS:

- resources,
- competence,
- awareness,
- communication, and
- documented information.

5.1 Resources (ISO 14001, Clause 7.1)

The leaders of the organization need to be ready to dedicate resources needed to establish, implement, maintain, and improve an effective EMS. In most cases it will not require large capital expenditures; initially the most significant resource outlay will be time.

5.1.1 Getting started

5.1.1.1 Taking stock of your current state

What resources are currently being allocated to environmental management activities? Are resources "hidden" in overhead budgets like maintenance, waste disposal, energy usage, regulatory reporting and staff training? If these resource expenditures can be segregated and tracked, the cost data can be used as one of the inputs to identify priority focus areas. It can also be used to demonstrate to top management bottom-line financial benefits of the EMS.

Evaluate the existing resources available within the organization, such as the quality or occupational health and safety management systems, or other functional groups. There may be ways to leverage EHS information systems, facility/maintenance capital expenditures, design engineering expertise, quality management auditors, or temporary staffing and internship programs to satisfy some of the EMS program resource needs.

5.1.1.2 What type of resources might be needed?

Consider the following:

- Time from internal staff to evaluate operations, collect data, attend training, develop EMS processes and coordinate EMS implementation.
- Specialized skills or services to evaluate environmental aspects or compliance obligations, respond to emergency situations, and perform audits.
- Information systems, to store and analyse monitoring and measurement data or maintain and control EMS documentation.
- Equipment and supplies to facilitate environmental operations such as waste/recycling sorting bins, spill clean-up supplies, secondary containment storage containers, computers, posters, etc.
- Engineered control technology for equipment and infrastructure.
- Financial resources for engineered control technology or external consultant support.

Resources are assessed during the management review to ensure they are adequate in light of future needs, such as new improvement plans, or changes in operations.

5.1.2 Helpful hints

Substantial environmental as well as financial benefits can result from implementing an EMS. Track the benefits as well as the capital and operational costs of environment-related activities so that the organisation can show cost savings (or cost avoidance) from pollution prevention initiatives, reductions in energy consumption and streamlined processes.

Be ready with the information needed to submit EMS-related funding requests in time for the annual budgeting process, or when funds come available at the end of the fiscal year.

Smaller organizations may have resource constraints but can tap the following external sources:

- Large customers, for technical information and assistance in EMS deployment.
- Industry associations and the local Chamber of Commerce, for information on common environmental aspects, sharing lessons learned and best practices, and joint engagement of external resources.
- Universities and other research centres, for information on control technologies, environmental conditions, and interns to assist in EMS deployment.
- Regulators, for regulatory information and compliance assistance.

Key linkages

- Leadership
- Planning actions to achieve objectives
- Management review

5.1.3 Check conformity

Requirement	Y	N	Follow-up actions
Adequate resources in the following areas have been identified and provided for establishing, implementing, maintaining and continually improving the EMS			
• knowledge and specialized skills			
• personnel time			
• equipment and supplies			

Requirement	Y	N	Follow-up actions
• control technology for equipment and facility infrastructure			
• information systems			
• financial			

5.2 Competence (ISO 14001, Clause 7.2)

Persons assigned a role in the EMS program need to be competent to perform that role. This includes employees, and even contractors, whose work can have a significant environmental impact or have an assigned responsibility within the EMS.

Competence is the ability to apply knowledge and skills. The question to ask is, "What knowledge, experience and skills related to our environmental issues are needed to perform the work?", and, "Do our workers have it, and apply it in their daily work practices?"

The competency required will vary for the different roles in the organization. Some workers, for example, may need to monitor wastewater processes for contaminants. If present, they must divert the effluent to a holding pond to prevent an adverse impact to the neighbouring waterway, and investigate the cause. They need to know how to operate the wastewater systems, have knowledge of potential contaminants being generated, and the permitted discharge limits. Other workers may need to have an in-depth knowledge of local and national environmental regulations, so they can correctly determine their applicability to the organization's operations, and advise the management team. Still others may need to be competent to perform an internal EMS audit, including knowledge of auditing practices, ISO 14001 requirements and environmental regulations, and experience conducting audits.

Knowing that competent persons are performing the EMS roles, or alternatively, knowing that additional skills, knowledge and/or experience are needed, is the first step to acquiring the competency needed.

5.2.1 Getting started

5.2.1.1 Taking stock of your current state

How do you know staff are qualified to perform their work activities? Often they are hired because they have met a set of criteria that the organization has defined for the job. Usually that criteria includes a certain amount of experience in their field, a level of education commensurate with the position, and/or skills that can be demonstrated. Alternatively, the organisation may have identified staff who show a willingness to take on more responsibility and have invested in their growth and development by providing additional training to them.

If there are job descriptions, they might be the first place to look for competency requirements; if not, the supervisors probably know the minimum skills and knowledge needed for the job. These sources may not have identified the environmental competency required for the job, but it may be a starting point.

Next, look at the existing formal and informal training programs. New employees may undergo orientation on company policies during their first days at work. Safety meetings may cover, for instance, chemical use, storage and disposal requirements. On-the-job training might be provided on work instructions or equipment operation and maintenance. Again, these may or may not cover all environmental issues, but can be augmented with the specific knowledge and skills that are appropriate for the job.

Lastly, compile information on the education, training and experience of workers.

5.2.1.2 Overview

Start by referring to the roles, responsibilities and authorities that were assigned (see Section 3.3). Then, knowing what skills and knowledge are necessary, determine the requisite education, experience and training needed for the various roles. If environmental regulations stipulate specific training requirements, be sure to include it. Also consider if a credential, such as degree, certification or license, is necessary.

Next determine whether the incumbents have the experience, education or training to satisfy the competence requirements for their job. Consider both formal and informal ways in which workers may satisfy the job requirements, such as schooling, employment history, apprenticeships, training courses, on-the-job training, conferences, or industry meetings. Take credit for skills and knowledge workers acquired in previous jobs.

In some cases, workers may not have the necessary and sufficient competencies so their skill set will need to be augmented. Where competencies are lacking, take action, either through additional training or mentoring of incumbents, hiring or contracting competent persons, or reassignment, to acquire the knowledge and skills needed.

NOTE Records are required to be retained as objective evidence of competency.

Lastly, if additional action was necessary to augment skills and knowledge, check that the action taken was effective. For example, if a worker receives additional mentoring or on-the-job training, observe his/her work practices to ensure the new skills and knowledge are being applied correctly. If a training course is given, include hands-on demonstration exercises or a proficiency test to verify the skills were acquired.

Records of worker competency are required. They can be captured in a list, spreadsheet or database. Records of formal education can include copies of diplomas, licenses, certificates or course work completed. Work history is typically found on the job application form. Training records usually include the date of attendance and course material. For example, on-the-job training is generally recorded on a check sheet listing the topics covered. Records of informal methods used will need to be developed internally. If contractors are used to fulfil a competency requirement, records of their competence are needed. Often they are submitted in the contract bid package, or provided upon request.

Key linkages

- Roles, responsibilities, authorities
- Environmental aspects
- Compliance obligations
- Resources
- Operational controls

5.2.2 Helpful hints

Enlist the human resources function. Incorporate the identification, evaluation and recordkeeping of EMS competency with the comparable processes for technical job competency. Efficiencies can be realized if the processes are integrated.

Grants or government funding may be available to help businesses train employees on environmental management.

A matrix can be used to summarize competence requirements, EMS related required training or licenses, and, if applicable, the requalification or retraining frequency for individuals or specific groups working for the organization. Table 9 provides a template that can be used.

Table 9 — Competency matrix template

Competence matrix template						
Target group	Education level	Experience level	Required training	Retraining frequency	Licenses or other qualification	Requalification frequency

5.2.3 Check conformity

Requirement	Y	N	Follow-up actions
Identify education requirements for EMS roles.			
Identify experience requirements for EMS roles.			
Identify training requirements for EMS roles.			
Verify workers satisfy competence requirements for EMS roles.			
Take action to acquire competence where gaps exists.			
Check that actions taken were effective in gaining the competence needed.			
Compile competence records (i.e. education, experience, and training).			

5.3 Awareness (ISO 14001, Clause 7.3)

It is important that workers embrace a proactive approach to protect the environment. This includes preventing pollution, complying with obligations and reducing environmental impacts on the job. An organization's awareness efforts facilitate development of a culture that reflects this proactive approach. These efforts have additional benefits in that they cultivate an environmental ethic within employees that are applied to activities outside the workplace. There are many techniques an organization can use to increase environmental awareness. Top management plays a key role in promoting an environmental culture.

Increasing environmental awareness

- Newsletters
- Signs and banners
- Earth Day celebrations
- Training programmes
- Staff meetings
- Mentoring
- Pamphlets
- Videos

5.3.1 Getting started

5.3.1.1 Taking stock of your current state

What methods are used to communicate important job information to employees? They may be formal or informal. For example, there may be formal training programmes for new employees such as safety training. Alternatively, important information may be communicated in more informal settings at periodic all-hands meeting, at daily "toolbox talks", or during lunch and breaks. All of these venues provide an opportunity to raise awareness on environmental issues.

How does the organisation communicate with contractors whose work has an environmental component? Typically a contract exists that establishes details of the work to be performed and terms and conditions, including relevant environmental requirements. How is it assured that the workers who show up at the work site actually know the details of the contract requirements? Perhaps a project manager assigned to the contractor conducts a pre-job briefing that covers, among other things, what to do with wastes and chemicals generated on the job.

5.3.1.2 What to cover

It is important that all persons working for or on behalf of the organization know how their work contributes to delivering improved environmental performance and making sure the EMS is effective. They need to be aware of the organization's environmental policy, and understand the significant environmental aspects associated with their work. Further, they need to be aware of the EMS requirements that apply to their job, and the potential consequences if operational controls developed for their job are not implemented, particularly those established to meet compliance obligations.

The requirements for competence, awareness and communication are complementary, so implementation of these requirements can be consolidated, or addressed using similar methods, thereby avoiding duplication of effort.

Awareness efforts do not have to cover all technical details (see Clause 7.2, Competence). Instead, awareness efforts should focus on the overall purpose, benefits and broad goals of the EMS. Employees that understand the big picture — the broad environmental challenges, how human activities impact the environment and the problems that are created — are motivated to address their environmental responsibilities proactively. The messages should cover:

- Why the organization is pursuing an EMS and what it hopes to accomplish (i.e. policy commitments and environmental objectives).
- How the organization minimizes adverse environmental aspects and impacts, and how the employee can help.
- The key EMS contacts within the organization, and their role and responsibility.
- The potential environmental and financial benefits to the organization.
- The value and importance the EMS program to the local community and other external interested parties.

Key linkages

- Environmental policy
- Environmental objectives
- Environmental aspects
- Compliance obligations
- Competence
- Communication
- Operational planning and control
- Emergency preparedness and response
- Nonconformity and corrective action

5.3.2 Helpful hints

One of the most effective ways to ensure employees are aware of their EMS responsibilities is to involve them in the EMS programme development and establish opportunities for them to provide ideas and feedback.

Awareness efforts are effective if they are delivered in simple, short messages that are repeated and reinforced over time or incorporated into a training course. They provide opportunities for getting feedback from employees on potential EMS programme improvements, so make sure to set aside time for interaction.

Workers don't need to memorize the environmental policy but they need to know it exists and the actions they take on the job that protect the environment, comply with requirements, or improve environmental management.

Just as new employee orientation covers the organization's mission, vision, values, objectives, rules and practices, it can also include information about the EMS policy, environmental objectives and the employee environmental responsibilities.

5.3.3 Check conformity

Requirement	Y	N	Follow-up actions
Mechanisms are in place to make employees and contractors working on site aware of their environmental management responsibilities.			
Employees and contractors are aware of:			
• the commitments in the environmental policy			
• the significant environmental aspects and impacts with their work			
• what they do on the job that contributes to the EMS' effectiveness and improved environmental performance			
• what can happen if they don't follow EMS requirements or don't comply with regulations			

5.4 Communication (ISO 14001, Clause 7.4)

Sharing information raises awareness, encourages partnerships and builds credibility. This forms the foundation for a successful EMS, both with employees and others who work with the organization. It has also become increasingly important to share information on the organization's commitments, actions and performance with external groups for these same reasons.

Communicating with employees and contractors is important to be sure the EMS requirements are understood. Internal communication facilitates coordination of activities, aids problem solving, and enables the organization to celebrate its progress.

Customers, suppliers, community members, regulatory agencies and investors all have an interest in the organization's practices and environmental performance. Proactive communications can be used to promote its accomplishments and influence opinions and misperceptions, building support and confidence in the business.

Depending on whether the organization wants to inform, consult with or involve the various target groups, there are multiple methods and channels for communicating environmental information both internally and externally.

Environmental communication principles

Transparency – making the processes, methods, data sources and assumptions available.

Appropriateness – making information relevant to the audience, using formats, language and media that meets their needs and enables them to understand and participate.

Credibility – provide truthful, accurate, substantive information in an honest and fair manner that does not mislead. Use recognized and reproducible methods and performance indicators.

Responsiveness – respond to queries and concerns in a full and timely manner, letting interested parties know how they were addressed.

Clarity – ensure communications are understandable and minimize ambiguity.

5.4.1 Getting started

5.4.1.1 Taking stock of your current state

Identify how current methods used to communicate with employees can be used for conveying environmental information. Consider staff meetings, training sessions and work instructions.

How does the organization communicate with suppliers and contractors? Would others methods be useful?

What obligations have to be communicated externally? Environmental laws and permits often include periodic reporting to regulators. Emergency situations typically have to be reported to local public agencies so that protective actions can be taken.

Voluntary reporting schemes, like the Carbon Disclosure Project or the Global Reporting Initiative, have specific information that needs to be reported on an annual basis.

Would there be value to the organization by communicating to other external parties? If the organization participates in community outreach activities, can they be used to promote its environmental initiatives? Are there environmental impacts associated with its products that should be communicated to external providers and customers?

How effective are the current methods of internal and external communication, and can they be improved?

5.4.1.2 Communication processes

Simply put, the organization's communication processes need to address the following:

- The key audiences, both internal and external — such as employees, suppliers/contractors, consumers, regulatory agencies and public agencies.
- What environmental information needs to be communicated to particular audiences — this will vary depending on their needs and how proactive the organization wants to be.
- The methods for communicating this information to the particular audience.
- The timing and frequency for communications, including prompt response to queries and complaints.

The organization can develop separate processes for internal and external communication or combine them into one. In either case, the processes developed start by using communication mechanisms currently in place. As the EMS matures and resources are available, the organization can expand and enhance its communications.

An approach that some organizations take is to capture their key communication activities in a matrix. Table 10 provides such an example.

Key linkages

- Scope of the EMS
- Leadership and commitment
- Environmental policy
- Roles, responsibilities and authorities
- Environmental aspects
- Environmental objectives
- Awareness
- Operational control
- Emergency preparedness and response
- Monitoring, measurement, analysis and evaluation
- Internal audit
- Management review

Table 10 — Communication matrix

XYZ Manufacturing environmental communication matrix					
Who	**What**	**When**	**How**	**Record**	**Notes**
Employees	Environmental policy and objectives; EMS responsibilities and requirements	Upon hire	New employee orientation		
		Annually	Annual environmental training program	Yes	
	Changes to EMS	As they take effect	Morning meetings		
			Intranet	Yes	
Management team	Progress on objectives	Quarterly	Operations meeting	Yes	
	Audit results	As they are reported	Email transmission	Yes	Legal counsel communicates results
	EMS performance	Quarterly	Operations meeting	Yes	
Contractors	Applicable EMS requirements	Upon contract execution	Contract terms and conditions	Yes	
		Before work begins	Onsite meeting with project manager		
Fire/police	Emergency preparedness information	Annually	Site tour		
	Emergency situations	When emergency occurs	Verbal notification — Call emergency number	Yes	Incident commander coordinates

XYZ Manufacturing environmental communication matrix					
Who	**What**	**When**	**How**	**Record**	**Notes**
Environmental agency	Air emission report	Quarterly	Electronic submittal	Yes	Environmental manager coordinates
	Toxic chemical inventory report	Annually, July 1	Electronic submittal	Yes	
	Emergency situations	Within 8 hrs	Verbal notification and written report	Yes	
	Noncompliance	When occurs	Verbal notification and written report	Yes	Environmental manager and legal counsel communicates
Carbon disclosure project	GHG emissions	Annually	Electronic submittal	Yes	
Community	Environmental policy, initiatives, accomplishments	Annually, coordinated with Earth Day activities	Posters/demonstrations		Employee volunteers
Customers	Environmental impacts; recycling and disposal options	Product development	Internet	Yes	
		Product delivery	Information sheet	Yes	

5.4.1.3 Environmental information to communicate

ISO 14001 requires a minimum amount of environmental information be communicated, reported or available to specific individuals and interested parties. Table 11 summarizes this information (but refer to the detailed ISO 14001 requirements in the clauses listed in the linkages box for more information). The organization, however, may provide additional information on its EMS as determined when developing its communication process.

Table 11 — Environmental information required to be communicated

Communication topics	Internal	External
Scope of the EMS		√
Environmental policy	√	√
Responsibilities/authorities for EMS roles	√	
Environmental objectives	√	√
Significant environmental aspects	√	√
Emergency preparations and response	√	√
Importance of effective environmental management and conforming to EMS requirements	√	
Changes to the EMS	√	
EMS performance	√	√ *
Audit results	√	
Responses to queries and concerns	√	√
Environmental requirements and impacts to external providers		√
Environmental impacts to users		√
* To the extent required by compliance obligations or identified in the organization's communication process.		

In all cases the information communicated should adhere to the environmental communications principles described above and be consistent with the information generated through the EMS to avoid omissions, misrepresentations, or errors.

5.4.1.4 Communication techniques

A single technique can meet the needs of many different interested parties, or multiple techniques might be necessary to address specific needs of individual parties. Most organizations tailor their environmental information for different audiences and make it available either on their internal and external websites. Environmental performance data can be summarized and reported on a periodic basis. Table 12 provides examples of communication techniques that an organization can consider using.

Table 12 — Communication techniques

Written communication techniques	Verbal communication techniques
Intranet/websites	Employee or public meetings
Environmental or sustainability reports	Interviews/focus groups
Brochures/newsletters	Informal discussions
Product labels or declarations	Press conference
Bulletin boards/posters/displays	Surveys
Letters/email	Site visits/guided tours
Training materials and work instructions	Presentations/workshops/ conferences
Media/news releases or newspaper articles	Employee committees, citizens advisory groups or community liaison groups

Since there is value in both providing and receiving information, two-way communication mechanisms need to be established. This enables interested parties to ask questions, make suggestions for improvement or express concerns about the organization's EMS or its environmental practices or performance. A combination of formal and informal methods work best. Whether an inquiry or comment comes from an internal or external source, it is important for the organization to respond promptly.

Formal, written communications should be retained, particularly external correspondence, environmental reports and regulatory submittals. It is also worth documenting the nature of environmental information provided verbally in public forums or at staff meetings. The organization can determine what additional communication it wants to keep records of, depending on whether it provides value. Summaries of communications, particularly complaints and concerns, can provide input for potential improvement.

5.4.2 Helpful hints

Tap into the persons who have responsibility and authority for official business communications. They have historical knowledge of inquiries made and techniques used. They can provide lessons learned for incorporation into the environmental communication processes.

Keep messages simple, clear, concise and accurate.

Top management may want to appoint a spokesperson for the organization when reporting legal compliance information, or dealing with public agencies during emergency situations.

Contract terms and conditions can be used to communicate environmental requirements to suppliers and contractors.

Warranties and information sheets can be used to communicate environmental impacts associated with product use, end of life treatment and disposal.

Benchmark other organizations to learn their environmental communication methods.

5.4.3 Check conformity

Requirement	Y	N	Follow-up actions
The organization has a process for internal communication on its EMS that covers: • what it will communicate • to whom it will communicate • the mechanisms it will use to communicate • frequency and timing of communications.			
The communication processes includes the mandatory and voluntary reporting of information required by its compliance obligations.			

Requirement	Y	N	Follow-up actions
Environmental information communicated is consistent with reliable information generated and reported within its EMS.			
The organization responds to inquiries and concerns expressed by internal and external parties.			
Two-way communication mechanisms are available for internal parties to make suggestions for improvement.			
The organization communicates the information required by ISO 14001, and any additional information it included in its communication process, internally.			
The organization communicates the information required by ISO 14001, and any additional information it included in its communication process, externally.			
Records of communiques are retained, as appropriate.			

5.5 Documented information (ISO 14001, Clause 7.5)

It is important for top management to ensure that EMS processes are carried out as planned and the desired results are achieved. They may also want to measure the speed of improvement, or record progress for an environmental report. Capturing key pieces of information in documented form can help in this effort.

Documenting how the system operates helps persons responsible for implementation understand what they need to do and how to do it. Where multiple persons are performing a process, documenting the steps can ensure consistency in the results. Documenting decisions made, EMS activities performed and the results achieved provides evidence to demonstrate conformity to requirements and effective implementation of the EMS. In addition, organizations may decide to document information for transparency purposes, continuity and historical perspective, new employee training or facilitating the audit process. Customers, regulators, lending institutions and certifying bodies may request environmental information, so having it in documented form will facilitate sharing this information.

Each organization needs to strike the right balance between implementing the EMS and maintaining documented information on the system. The focus should be on ensuring activities are carried out as planned so they meet the EMS requirements, not on a complex documentation system. In some cases, small organizations may be able to accomplish this goal using communication and training methods, rather than documented information.

There are basically two types of documented information; "living" documents that describe how things are done within the EMS, and "static" records that reflect results of some activity at a point in time. Whether electronic or paper, the correct and current versions of living documents such as plans, work instructions and programmes need to be readily available to those who use them. This requires the organization to have a mechanism to create these documents and control their revision. Records of results need to be created, reviewed and retained for a period of time. When superseded or no longer of value, documents and records need to be dispositioned. The organization may want to retain a copy for historical purposes, but it is important the old versions are removed from the workplace so they are not inadvertently used.

5.5.1 Getting started

5.5.1.1 Taking stock of your current state

Look at how the organization maintains its business-related documents. Are they stored electronically in a central location, or is it in paper form maintained in separate files by function? Is the environmental-related documentation stored similarly? Probably there is a mix, so an evaluation of what works well, and what doesn't, can be used to streamline the way EMS documentation is managed.

Before creating new documents, look for documents already created for other management systems that can be used to satisfy EMS requirements. Examples include using the nonconformity and corrective action documentation process from the quality system for tracking EMS nonconformities, or using the human resource training and qualification system for tracking required environmental training.

If an organization has implemented a quality management system, there is probably a process already in place that satisfies most, if not all, EMS requirements for documented information.

5.5.1.2 Document management

The organization needs to determine what information needs to be documented, controlled and maintained so those responsible for implementation understand what they need to do to satisfy EMS requirements. This ensures the EMS processes are carried out as planned and consistent results are produced. As the EMS develops, processes may change with the introduction of new equipment, or discrepancies in operational procedures may be identified. In these cases, documented instructions will need to be or amended. An environmental management system is a "living" system and an organization needs to track the changes it makes. Where ISO 14001 states "documented information shall be maintained", it means a document.

There are certain documents that are required by ISO 14001 (see Table 13), but the organization determines if other documents are needed to ensure the effective planning and operation of EMS processes. There may be sector-specific codes of practice, equipment manuals or regulatory permits/licenses generated by external sources that are important to the way operations are carried out. Copies of these external documents need to be managed as well.

When determining what, if any, additional information should be documented, consider the size and culture of the organization. Small organizations may work effectively using less formal methods, such as communicating required information verbally or providing it during staff meetings or training classes. Larger organizations, with many persons involved in EMS implementation may find it more effective to provide written information in the form of work instructions or programme descriptions so that everyone has the same information. Whether the organization is small or large, a complex process that requires coordination to work properly may be easier to execute if documented. Another consideration is whether the organization needs to formally demonstrate to external parties, such as regulatory agencies, customers, or certification bodies, evidence of conformance to requirements.

The following points need to be addressed to ensure documents are managed in a controlled manner:

Preparation — Provide some unique identifier (e.g. ID number and date) and description (e.g. title). Decide the format (e.g. language, visual graphic) and media (e.g. electronic, paper). Establish a review and approval mechanism, including who has approval authority.

Issuance/distribution — Distribute to those who need it in a format they can access readily when and where they need it. (If employees don't have access to computers in their work area, electronic versions of documents are not readily accessible.) Keep track of where the document is distributed so that updates can be re-issued when available.

Revision — As changes to processes occur, the documented information may have to be revised to remain accurate. Basically the points that apply to preparing a new document (i.e. unique identifier/description, format, review and approval) and issuing/distributing apply to a revision. What is most important is to replace old documents with the revised ones so that everyone is working with the correct information. It may be prudent to keep one as a "historical" record.

Table 13 — **Documented information required by ISO 14001**

Documents	Records
• scope of the EMS • environmental policy • risks and opportunities • process(es) to satisfy Clause 6.1 * • environmental aspects and impacts • criteria for significant environmental aspects • significant environmental aspects • compliance obligations • environmental objectives • operational controls * • emergency preparedness and response *	• competence • internal and external communications, as appropriate • monitoring, measurement, analysis and evaluation results, as appropriate • compliance evaluation results • audit programme implementation • audit results • management review results • nonconformities and actions taken • corrective action results
* Each organization decides the extent to which documented information is needed to have confidence that these processes are carried out as planned.	

Key linkages

- EMS scope
- Environmental policy
- Risks and opportunities
- Environmental aspects
- Compliance obligations
- Environmental objectives
- Competence
- Communication
- Operational control
- Emergency preparedness and response
- Monitoring, measurement, analysis and evaluation
- Compliance evaluation
- Internal audit
- Management review
- Nonconformity and corrective action

5.5.1.3 Records management

An organization needs to determine information needed to provide evidence that EMS activities are performed and the results achieved. These are called records. Where ISO 14001 states "documented information shall be retained", it means a record. Records needs to be documented, retained and controlled. This ensures an organization can demonstrate conformity and effective implementation of the EMS. Records also provide evidence of the "dynamic" nature of the system. Where new materials are introduced or new equipment installed, it is important to be able to trace results to those changes. A dated record of implementation of such changes should be retained for reference and to support data analysis and evaluation processes.

There are certain records that are required by ISO 14001 (see Table 13), but the organization determines if other records are needed. Virtually every EMS element can result in a record, but the key is identifying what records are needed to evaluate EMS implementation. Also consider the need to formally demonstrate evidence of conformance to ISO 14001 or requirements to external parties, such as regulatory agencies, customers, or certification bodies. Contractors and other external suppliers may generate records that the organization needs, so it is important to establish mechanisms to retain copies of these or have ready access to them.

The following points need to be addressed to ensure records are managed in a controlled manner:

Generation — Provide some unique identifier (e.g. name of person creating record, date) and description that provides traceability to the activity (e.g. location, title, form number). Decide the format (e.g. language, visual graphic) and media (e.g. electronic, paper). Establish a review mechanism, including responsibility. Records are not revised per se, but may undergo correction.

Storage — Protect or preserve the records to retain legibility and prevent damage, loss and unauthorized access. Organize them so that they are retrievable (e.g. file by date, or create an index). Backup electronic records, and be sure the software required to read them is stored and protected for future use.

Retention — Identify the period of time records need to be retained. Take into account compliance obligations, such as contracts and legal requirements that impose specific retention periods. Dispose of records in accordance with the retention schedule.

Table 14 provides a template for managing EMS records.

Table 14 — Records management template

Records management template					
Record description	Owner	Format/ media	Location	Access/ distribution	Retention period

5.5.2 Helpful hints

Documents that are simple and easy-to-read are easily understood and implemented. Flowcharting is one way to document a process that shows the steps visually.

A matrix showing the list of EMS documents (and latest revision) and who/where they are needed can be used to control distribution.

Storing documents on an internal website or local area network makes updating and distributing information simple as long as those who need them have computer access.

When revising documents, provide a summary description of the key changes so that users can find them easily.

An organization may find it useful to document a summary description of the main elements of its EMS and how they fit together. This can be documented in a manual and include references to related documented processes rather than duplicating the information.

EMS documentation can be controlled using existing information management systems. Incorporating environmental monitoring and measuring data into electronic database systems, and maintaining retention schedules in the organization's records management system, are some examples.

For organizations pursuing ISO 14001 certification, some documents and records should be retained for the three year certification cycle at a minimum, such as management review records.

EMS audits verify the organization is retaining required records, and often checks that they are being disposed in conformance with their retention schedule.

5.5.3 Check conformity

Requirement	Y	N	Follow-up actions
Information required by ISO 14001 to be documented is maintained.			
Evidence of results required by ISO 14001 to be documented is retained.			
Other information required by the organization to be documented is identified and maintained/retained, including information generated by external sources.			
The creation and updating of documented information includes: • Identification and description. • Format and media. • Review and approval.			

Requirement	Y	N	Follow-up actions
Controls for ensuring documented information is available to users and protected are in place, including: • Distribution. • Access. • Storage and preservation. • Retrieval. • Change control. • Retention. • Disposition.			

6 Implementing operational controls

Establishing, implementing and maintaining control of the organization's operations is critical to ensuring that EMS requirements are met. It is part of the component of the Do phase of the Plan-Do-Check-Act model. ISO 14001 identifies two clauses to assist an organization in this effort:

- Operational planning and control, and
- Emergency preparedness and response.

6.1 Operational controls (ISO 14001, Clause 8.1)

Operational controls are one of the important ways environmental impacts and other adverse effects are prevented. Controls are also used to assure compliance with obligations and to implement improvements. They ensure that day-to-day activities are performed consistent with established plans and objectives. Operational controls are put in place to ultimately ensure the environmental policy commitments are achieved.

Operational controls specify how business processes are executed, including the allowable acceptance criteria. Some operations are controlled using automatic engineered technology, such as an energy management system that regulates heating, ventilation and air conditioning levels to predetermined set points. Others are controlled by specifying design criteria, using documented procedures or posted warnings, or by monitoring key characteristics. Not only can they be used to control processes within the organization's fence line, they can be used to influence upstream and downstream processes in the product life cycle. The key is identifying the most effective control points in the process and implementing efficient, yet cost effective controls.

6.1.1 Getting started

6.1.1.1 Taking stock of your current state

Are there existing programmes or processes to control significant environmental aspects or to ensure regulatory compliance?

Review the operations and activities that have significant environmental aspects and compliance obligations. Inventory the operational controls currently in place. Are they adequate to controlling important process steps? If not, consider upgrading controls.

6.1.1.2 Overview

Organizations need to establish, implement, control and maintain process(es) to:

- meet EMS requirements;
- implement actions planned for significant environmental aspects;
- implement actions planned to satisfy compliance obligations;
- implement actions planned for risks and opportunities to achieving the intended outcomes of its EMS;
- implement actions planned for achieving environmental objectives.

The processes established need to:

- stipulate operating criteria for the affected activities, products and services,
- have adequate control(s) to ensure the processes are implemented in accordance with the operating criteria.

Not only does control need to be applied to stable operations, it also needs to be applied when operations are undergoing change. Control needs to be maintained while planned changes are put in place, making modification to the operating criteria and controls as needed. For example, if equipment is being replaced, there may be a period of time when engineered controls are not operational. During this period, temporary changes to operations may need to be made. Planned changes may have unanticipated consequences. In these cases, an organization needs to mitigate adverse environmental impacts if necessary.

Documented information of the control processes that are implemented need to be maintained up-to-date. The extent of documentation depends on what is needed by the organization to ensure the controls operate as intended.

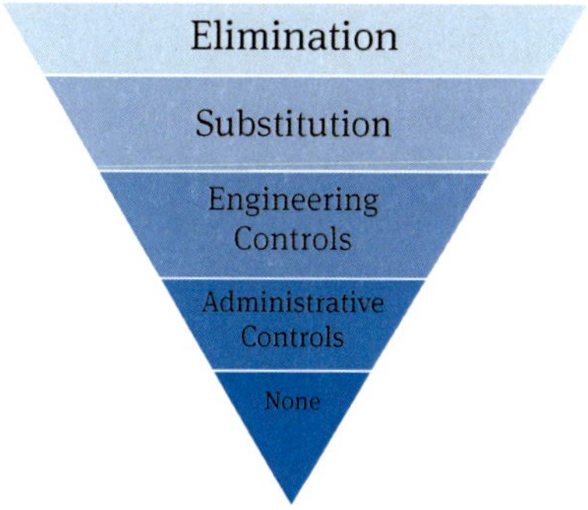

Figure 7 — Hierarchy of control

Controls can take many forms, and can be implemented singly or in combination. It is preferable to follow the hierarchy of controls when selecting controls, or evaluating the efficacy of current controls (see Figure 7).

A description of the concepts, as applied to significant environmental aspects is provided below. The same concepts can be applied to the control of processes needed to ensure compliance obligations are satisfied and to address risks and opportunities.

Elimination — Eliminating the significant environmental aspect, or eliminating the operation, activity, material or service that creates a significant environmental aspect, is the most effective control.

Substitution — The second most effective control, involves replacing something that produces a significant environmental aspect (similar to elimination) with something that does not produce one, or produces a less significant one.

Engineering controls — The third most effective means of controlling processes is engineered controls. These do not eliminate environmental aspects, but rather isolates it and prevents it from escaping into the environment. Dust collectors and neutralization systems are examples, as is error-proofing a process.

Administrative controls – Administrative controls do not remove or address the significant environmental aspect directly, but instead control the way people work with it, so as to limit or prevent its interaction with the environment. Examples of administrative controls include work instructions, training, warning signs and monitoring/measurement techniques with associated record keeping. Administrative controls are the least reliable for preventing an environmental impact, but may be adequate to control less complex processes.

If no operational controls can be identified, the significant environmental aspects is insufficiently addressed.

Key linkages

- Environmental aspects
- Compliance obligations
- Risks and opportunities
- Planning action
- Environmental objectives
- Communication

6.1.1.3 Environmental requirements for design

When environmental considerations are taken into account early in the process, product or service development, potential environmental impacts at each stage of the product life cycle can be avoided or controlled. Similarly, the associated costs of controlling these impacts, particularly those associated with end of life treatment and disposal, can also be minimized. Many have learned through their quality program that product quality attributes are dependent on decisions made at the design stage; so too are environmental attributes. Rather than rely on "end of pipe" controls and treatment, an organization can use their design and development processes to protect the environment.

NOTE The operational controls are required to be documented and kept up-to-date.

An organization needs to establish a process whereby environmental requirements for each life cycle stage are addressed in the design and development of products and services, including:

- acquisition of raw materials,
- production or manufacturing processes,
- transport of the product to market or delivery of service to the customer,
- use of the product or service,
- end-of-life treatment and final disposal.

By focusing on the design and development process, organizations can

- understand the environmental aspects associated with inputs and outputs of materials and by-products at each stage of a process, product or service life cycle;
- incorporate environmental specifications or requirements for inputs and outputs at the optimal points along the life cycle;
- seek out alternatives that:
  - avoid use of banned chemicals,
  - substitute less hazardous materials,
  - minimize natural resources and energy required during product use,
  - maximize use of recyclable or reusable packaging materials,
  - enable product (or component) reuse or recycling at end of life;
- apply lessons learned to design or redesign of other processes, products and services.

6.1.1.4 Control of external suppliers, contractors and outsourced processes

When deciding controls for goods and services provided by external entities, a clear understanding of the degree of control the organization can and should exercise is necessary, as well as recognition of limitations which would dictate when it has little or no influence.

The organization typically has direct control over its procurement of goods and services. Environmental requirements established in the organization's design and development processes can be communicated to the respective suppliers of materials. These requirements are typically reflected in the particular choice of goods procured or elaborated in contract terms and conditions.

ISO 14001 requires that the organization's environmental requirements be communicated, where applicable, to contractors and service providers. It is easy to understand why it is important, knowing how frequently contractors and suppliers visit a facility to undertake maintenance or other services. They access energy and water supplies, utilise waste management facilities and, unless suitably briefed on the organization's environmental requirements, can cause pollution, bring unacceptable materials onsite, and waste energy or other resources. They deliver products to market on behalf of the organization. There could be environmental impacts associated with transport of products or delivery of services that these external providers need to know. This information is often communicated through shipping papers, markings on containers and appropriate placarding on the fleet vehicles used in transport.

Applicable environmental requirements established in the organization's design and development process, as well as any additional controls to minimize environmental damage, can be specified within contract terms and conditions. Work instructions used by external providers can be requested and reviewed for acceptability. Alternatively, work instructions can be provided and then reinforced when the contractor's representatives arrive. Lastly, work performed onsite can be controlled through oversight activities.

However, typically an organization will seek to use an external provider of products and services when those products or services exceed its own capabilities or are outside its core mission. In some cases, the organization is relying on the knowledge, competence and resources of the external provider to know how best to deal with environmental aspects of the work. In these cases, during its selection process, organizations can seek from its potential vendors information on how the service provider will address its environmental aspects and recommendations for limiting environmental impacts. This can help an organization select the best value service provider.

Outsourced processes is an arrangement where an external organization performs part of an organization's function. Similar to externally provided services, the organization may choose to outsource because they lack the equipment, capability and competence or they do not wish to do it in-house. In any case, the organization should and can rely on the external provider to control the environmental impacts from the outsourced process. Even though an organization may have limited or no influence on the provider of the outsourced process, there could be cases where they retain liability for conforming to requirements, explicitly or implicitly, particularly if the relationship is such that interested parties perceive the process being carried out by the organization. As such, organizations have a responsibility to exercise due diligence to be sure the external provider is operating the particular outsourced process in an environmentally sound manner. The organization needs to communicate its expectations and requirements for the process, and may also request work instructions, records and audit results from the outsourced entity that demonstrate it has met the requirements. The type and extent of control to outsourced processes shall be defined within the environmental management system.

In all cases of externally provided goods and services, the level of control an organization exerts is commensurate with the environmental aspects, risk and opportunities, and compliance obligations.

Communicating

Environmental requirements to suppliers and contractors of goods and services.

Information on environmental impacts to:

- external providers associated with transport of products, or delivery services, to market;
- consumers, associated with use or end-of-life treatment or final disposal of products.

6.1.1.5 User/consumer information

Some of the environmental impacts associated with an organization's products or services can occur during the use, end of life treatment or final disposal stages. By providing information to customers and users/consumers on environmental impacts associated with products and services, and preferred options for disposal at end of its useful life, these impacts can be avoided or minimized.

6.1.2 Helpful hints

The organization's commitment to improving its environmental performance should be communicated beyond employees, to suppliers and contractors. They can provide useful recommendations and best practices for controlling processes.

There is increasing pressures on businesses from external stakeholders to define and exercise the necessary and sufficient controls on outsourced processes. This requires careful examination and in some cases, the use of experts to assist the organization in defining controls or assessing the outsourced process.

6.1.3 Check conformity

Requirement	Y	N	Follow-up actions
The organization establishes, implements controls and maintains processes to:			
• meet EMS requirements,			
• implement actions planned for significant environmental aspects,			
• implement actions planned to satisfy compliance obligations,			
• implement actions planned for risks and opportunities to the organization and its EMS,			
• implement actions planned for achieving environmental objectives.			
The processes established need to:			
• stipulate operating criteria for the affected activities, products and services,			

Requirement	Y	N	Follow-up actions
• have adequate control(s) to ensure the processes are implemented in accordance with the operating criteria.			
Planned changes are controlled and consequences of unintended changes are reviewed. Action to mitigate any subsequent adverse environmental impacts is taken as necessary.			
Outsourced processes or influences are controlled, with the type and extent applied being defined.			
The organization shall establish controls to ensure environmental requirements are addressed in the product (or service) design and development process for the relevant life cycle stages.			
Environmental requirements are determined and communicated to external providers (suppliers and contractors) for the procurement of products and services.			
Communicate potential significant environmental impacts as appropriate, associated with			
• transportation of products or delivery of services			
• product use			
• end-of-life treatment and final disposal of products.			
The organization maintains documents on process controls to the extent needed to be sure the controls are implemented as required.			

6.2 Emergency preparedness and response (ISO 14001, Clause 8.2)

Unplanned or unexpected events require an immediate response to minimize adverse impacts to the environment. In addition, the organizational effects can be expensive and damaging — downtime, property loss, injuries, environmental clean-up costs. Further, emergency events can erode confidence and damage the organization's reputation.

Emergency situations can originate internally or externally. Waiting until the emergency situation occurs, puts the organization in a reactive mode, which is often inefficient and ineffective. However, if an organization anticipates and plans for reasonably foreseeable emergency situations, it can take a proactive stance to prepare ahead of time, taking action to prevent emergency situations from occurring where practicable, and if not, the organization can coordinate with local response agencies, train their employees, and mitigate the adverse environmental impact when it does occur.

6.2.1 Getting started

6.2.1.1 Taking stock of your current state

Most organizations have an emergency plan to protect the health and safety of employees and prevent property loss. Typically this addresses what to do in the case of fire, at a minimum, and includes an evacuation plan and periodic fire drills. This is a good starting point for identifying emergency situations that can have environmental impacts and the preparations needed to respond.

6.2.1.2 Overview

There are several requirements in ISO 14001 that pertain to emergency situations:

- identification of potential emergency situations (Clause 6.1.1)
- establishment, implementation and maintenance of a process to prepare for and respond to potential emergency situations (Clause 8.2). For ease in use, both are described in this section of the guide.

A cross-functional team made up of facilities/maintenance, operations, and environment, safety and health personnel is best suited to identify potential emergency situations. They understand the plant layout, the location of hazardous materials, normal and abnormal (start-up, shutdown, maintenance) operations, safety systems and local codes. Identification of potential receptors — i.e. land, waterways, flora, protected species, and nearby communities — is also important. It enables the organization to focus its efforts on preventing and/or planning a response to mitigate the specific receptors that could be adversely affected by an emergency situation.

Some examples of emergency situations are:

- Unsecured fuel oil tanks are dislodged during a natural disaster, contaminating the property.
- Run-off of fire water contaminated with hazardous materials causes adverse impacts to the groundwater and nearby surface waters.
- Start-up operations for a new piece of equipment results in accidental release of hazardous air pollutants.

Potential emergency situations

- Fires
- Explosions
- Petroleum or chemical spillages
- Construction failures
- Floods
- Earthquakes
- Tornados
- Lightning strikes
- Uncontrolled emissions to air, water or land
- Power failures
- Sabotage or terrorist threats

When preparing for emergencies, an organization can apply the hierarchy of controls. To the extent that actions can be taken to eliminate the impact in the first place, they should. This may involve securing outdoor fuel tanks and taking protective measures to contain contents if lines break. In other cases, being prepared to respond can be adequate. The organization can install emergency spill kits containing equipment to control the accidental release of potential pollutants. It can establish processes for operation, periodic inspection and maintenance. By training workers on its use and conducting mock drills, staff become experienced in operating the devices.

NOTE The emergency preparedness and response plan is required to be documented and kept up-to-date, incorporating lessons learned from emergency response and drills.

A critical component of emergency preparation is training staff on actions they are to take when an emergency occurs. Consider how to provide contractors and visitors with emergency response protocols. Evacuation routes need to be planned, clearly marked and free of obstructions. Local officials and fire brigades also need to prepare for emergency responses. As such, they welcome opportunities to tour buildings and participate in drills. These encounters provide opportunities to convey the organization's commitment to protect the environment and seek input on how to prevent adverse impacts.

What is most important when an emergency occurs is to call the experts. Therefore it is necessary to compile emergency numbers for notifying public emergency responders, hospitals, regulators, clean-up services and key personnel. Assign responsibility to key personnel to assist the agency responders in their efforts to mitigate the environmental impacts. Consider appointing a company spokesperson.

ISO 14001 requires the planned response actions to be tested periodically. Drills are ideal opportunities to test emergency preparedness. They can highlight deficiencies in the location and labelling of emergency equipment or escape routes, for example. Lessons learned from these tests can be used to update or improve the protective measures and procedures in place.

The emergency preparedness and response plan needs to be documented. The level of detail and extent of documentation depends on what is necessary to ensure the plan is executed as planned. The plans need to be reviewed periodically, particularly after tests or actual emergencies occur, and as necessary, revised.

Communicating

- Emergency situations that result in releases are reported to local agencies so that protective actions can be taken.
- In some cases, where the release exceeds regulatory thresholds, the environmental regulatory authority is notified.

Key linkages

- Environmental aspects
- Risks and opportunities
- Competence
- Awareness

6.2.2 Helpful hints

Consider the potential emergency situations that can occur from environmental conditions such as flooding due to eroded marshlands, extraordinary weather conditions related to global warming, and other natural disasters.

Decide what response actions the organization wants its personnel to engage in. If there is a fire, are they expected to operate a fire extinguisher and put out the fire, or notify the public agency and evacuate? Training and competency of personnel needs to be commensurate with the response actions expected.

6.2.3 Check conformity

Requirement	Y	N	Follow-up actions
Establish, implement and maintain an emergency preparedness and response plan to address emergency situations that have the potential for adverse environmental impact.			
Plan actions to prevent or mitigate adverse environmental impacts from emergency situations.			
Respond to emergency situations taking action to prevent or mitigate adverse environmental impacts.			
Conduct drills, or exercise the response plans.			
Review plan periodically, including after emergency situations and drills/ exercises, and revise as necessary.			
Provide information and training to relevant internal and external interested parties (employees, contractors, public emergency personnel, for example).			
Document the emergency preparedness and response plan.			

7 Evaluating EMS performance

Assessing how well the EMS is performing is a critical component of the check phase of the Plan-Do-Check-Act model. ISO 14001 contains several clauses that address assessing performance:

- Monitoring, measurement, analysis and evaluation
- Evaluation of compliance
- Internal audit
- Management review.

7.1 Monitoring, measurement, analysis and evaluation (ISO 14001, Clause 9.1.1)

Monitoring and measurement is about collecting reliable data on the performance of the EMS and the results that are being achieved. Analysis and evaluation is about using that data to make objective judgements on whether the EMS is operating as planned and whether the results are as good as anticipated. Just as the old adage goes, "If you can't measure it, you can't manage it", collecting and analysing environmental performance information is a critical component of managing the organization's significant environmental aspects. The information is used to confirm where the system is working well and producing acceptable results, and where it is not. It is an important tool used to make decisions about where to invest resources to improve.

7.1.1 Getting started

7.1.1.1 Taking stock of your current state

Make a list of the monitoring or measurements currently being performed related to control technology equipment, regulatory requirements, or efficiency measures (i.e. energy, material use, wastes).

Review the data collection and information management systems currently used for environmental data and those used by other functional units. The group responsible for facilities maintenance may be controlling the heating, ventilation and air conditioning using an energy management system. The accounts payable department may be collecting information on water and energy bills. Human Resources may have a system for tracking staff qualifications.

7.1.1.2 Monitoring and measurement

There is a subtle but important difference between monitoring and measurement. Monitoring is determining the status of a system, a process or an activity, which might entail a need to check, supervise or critically observe. This term is generally used when referring to a series of observations made over time without necessarily using measuring equipment. On the other hand, measurement is a set of activities performed to determine a value. Measurement generally refers to using some type of equipment to determine quantitative or qualitative properties. Some examples are shown in Table 15. But keep in mind there are no hard and fast rules on what has to be monitored versus what has to be measured; often a single process can be either monitored or measured, or both. As an example, the negative pressure required to minimize air emissions from a paint booth in an automobile body shop can be "monitored" by verifying that the pressure reading is within the acceptable range, or can be "measured" by creating a daily log of readings from the Magnehelic gauge. What is important to note is that sometimes it is appropriate and sufficient to observe whereas other times it is necessary to take measurements.

Table 15 — Examples of monitoring and measurement

Monitoring	Measurement
Progress on action plans to achieve environmental objectives	BOD, COD, TSS and pH in waste water discharges to fulfil a regulatory requirement
Timely implementation of corrective actions	CO_2 equivalent from burning fossil fuels
Completion of training requirements	Boiler NOX, SO_2, CO, and TSP emissions
Biodiversity survey (counting number of species per m^2)	Year over year reductions in solid waste sent to landfill
Employee involvement in community outreach activities	Smoke opacity from diesel truck
Checking for leaking or full waste containers	
Checking operating parameters of air scrubber are within acceptance limits	

The topics that need to be considered when determining what to monitor and measure include:

- significant environmental aspects,
- fulfilment of compliance obligations,
- environmental objectives,
- operational controls.

Specific guidance on monitoring, measuring, analysing and evaluating the organization's adherence to its compliance obligations is given in Section 7.2.

Initially it may suffice to monitor and measure the final outcome (i.e. total quantity of air emissions; number of violations of regulatory requirements; achievement of the objectives; percentage time equipment operates within established criteria).

Additional data may need to be collected on the characteristics associated with the activities and processes that contributed to the outcomes. This data can be used to confirm adherence to requirements, analyse performance problems or pinpoint areas requiring additional control.

Following earlier examples, the automobile body shop might measure the quantity of paint used per hour as a predictor of air emissions. An organization finding itself missing important regulatory reporting deadlines may want to monitor the cycle time for processing regulatory reports. To ensure achievement of environmental objectives top management can monitor indicators of progress quarterly. Collection of post-treatment wastewater samples from a plating process may be needed to confirm heavy metal concentrations before discharge.

The challenge is to identify the critical few characteristics that are the key contributors to an outcome that the organisation wants to achieve. Once identified, the monitoring and measurement methods and frequency for their conduct need to be defined.

To ensure valid quantitative results the methods used need to be capable of generating reproducible, traceable and reliable results. The methods need to be conducted under controlled conditions, such as:

- selecting established and industry accepted sampling and data collection techniques;
- calibrating or verifying measuring equipment using traceable measurement standards and appropriate quality control techniques;
- using competent personnel.

Instrumentation, engineered control technology and data collection systems are becoming more automated. Remote, often wireless-connected, equipment is increasingly used to collect samples, monitor operations or capture data such as emissions to air and water. This equipment requires calibration and maintenance, for which records should be kept, to ensure the data is accurate.

When considering the frequency, coordinate the timing of monitoring and measurement with internal schedules for analysis and evaluation results so that data is available when needed. Also take into account any reporting deadlines required by compliance obligations.

NOTE Records are required to be retained as objective evidence of monitoring, measurement, analysis and evaluation results.

Consider including monitoring logs, sampling plans, laboratory analytical results, data summaries and trends analysis.

Communicating environmental performance

- Monitoring/measurement data and information to managers of units where it was collected
- Summaries of analysis and trends to top management
- Analysis results to regulators if required
- Summaries of analysis and trends to others required by a compliance obligation or identified in communication process

7.1.1.3 Analysis and evaluation

The data generated through either monitoring or measurement is transformed via summarization, trending and, in some cases, comparison to peers, into information that can be used by top management to make decisions about conformance to requirements, the effectiveness of the EMS and the organization's environmental performance.

It is helpful to express the data using performance indicators that are simple to generate, relevant, and understandable to those who need to act on the information. The indicators used may, and should, vary for the audience. Energy use per unit of production is actionable by the operations supervisor, but energy spend per quarter gets the attention of the CFO. Metric tonnes of CO_2 emitted per year is required to be reported by organizations voluntarily reporting to the Carbon Disclosure Project and other greenhouse gas protocols. But workers who are involved in greenhouse gas reductions initiatives may find it easier to understand the percentage of the goal achieved to date.

Analysis of monitoring and measurement data can identify trends and relationships. A comparative analysis to previous performance over time indicates how much improvement has been realized by the actions taken. Cause and effect relationships can be revealed that point to areas where small changes can yield positive results. Benchmarking to external sources such as peers and best-in-class performers in similar industries can give insight on how the organization compares to its competitors. The analysis methods selected should be simple and sound. Exercise care when employing statistical techniques such as hypothesis testing and regression analysis to be sure they are applied correctly.

Use established criteria for evaluating data. Criteria to judge progress on objectives is inherent in the timelines and the overall change expressed in the objective (i.e. a reduction in pollutant level; an increase in recycled paper used in printers). Operating criteria established for processes needing operational control can be used to evaluate how consistently the process met its requirement. An internal administrative limit for wastewater discharges, set at 75 % of the regulatory permit discharge limit for example, not only provides criteria for evaluation but provides a "safety net" that ensures zero permit excursions.

Analysis and evaluation is not limited to monitoring and measurement data. The results from evaluation of compliance (Section 7.2) and Internal Audit (Section 7.3) also undergo analysis and evaluation. The timing of all three of these analyses and evaluation activities needs to be coordinated with the management review cycle so that top management has the information it needs to judge the suitability, adequacy and effectiveness of the EMS.

Key linkages

- Environmental aspects
- Compliance obligations
- Objectives
- Communications
- Operational control
- Management review

7.1.2 Helpful hints

This is a resource intensive activity, so be sure that there is a purpose for each piece of data that is being collected, and that it is being actively used in decision-making.

Don't overdesign the data collection and information management systems needed. Decide what is sufficient to be visually checked, versus jotted down on a weekly log, versus entered and tracked in a spreadsheet, versus compiled, tracked, analysed using a custom-tailored database management system.

7.1.3 Check conformity

Requirement	Y	N	Follow-up actions
The organization has determined what needs to be monitored or measured, including the following:			
• Significant environmental impact.			
• Compliance obligations.			
• Progress on objectives.			
• Operational controls.			
Methods for monitoring and measurement are established.			
The frequency for conducting monitoring and measurement activities is established.			
Calibrated or verified equipment is used for measurements.			
Methods for analysing monitoring and measurement data are established.			
The criteria for evaluating monitoring and measurement data is established.			
The frequency of analysis and evaluation is established.			
Environmental performance and EMS effectiveness is evaluated.			
The results of monitoring, measurement, analysis and evaluation are documented.			
Environmental performance information is communicated internally, as identified in its communication process.			

Requirement	Y	N	Follow-up actions
Environmental performance information is communicated externally, as required by the organization's compliance obligations and to other external interested parties identified in its communication process.			

7.2 Evaluating compliance (ISO 14001, Clause 9.1.2)

Organizations are required by law to comply with relevant environmental legislation. In addition, under certain circumstances, regulatory authorities set performance limits for compliance, which are specific to the organization's activities. An organization may also have other types of non-legal requirements to which they must adhere. These may be codes of practice relevant to its industrial sector, or contractual requirements in relation to products or services provided. There are ramifications for organizations that do not fulfil their requirements that can range from damage to the company brand, contract termination, citations of noncompliance, fines and even jail time, depending on the severity of the infraction and the type of requirement. Therefore it is important that an organization has a robust process to self-assess compliance to its requirements, self-identify non-compliant situations and self-correct non-compliances in a timely fashion. The results of these evaluations need to be reported to top management so they can maintain knowledge of the organization's compliance status at all times. The information on noncompliant situations provides an important input to prioritizing improvement.

The analysis described in Section 4.2 provides the details of these requirements and how they apply to an organization. This section describes how an organization evaluates its adherence to its compliance obligations.

Communicating

Results of compliance evaluations shall be reported to top management during the management review.

7.2.1 Getting started

7.2.1.1 Taking stock of your current state

Use the list of compliance obligations generated earlier (Clause 6.1.3) to determine the requirements which must be evaluated.

What techniques are conducted routinely to evaluate compliance with environmental legal requirements and permit conditions? Reviewing analytical results of wastewater discharges samples before submittal to the environmental agency is a form of evaluating compliance, just as weekly inspections of hazardous waste storage areas are.

Compile information on the scope and schedule of environmental-related audits and inspections performed by external parties on the organization, such as regulatory agencies, customers, or other third parties.

Review the processes used by other functional units in the organization to monitor regulatory compliance, such as safety, export and import, and contract law. Can these processes be applied to the evaluation of environmental compliance obligations?

7.2.1.2 Overview

Consistent with an organization's commitment to fulfil its compliance obligations, the organization has to establish, implement, and maintain a process for periodically evaluating compliance with applicable requirements, including regulatory requirements, customer requirements and other voluntary environmental-related requirements that it has undertaken. These evaluations include monitoring of facility activities, inspecting work areas and equipment and checking worker adherence to operating criteria. Notation of malfunctions, deterioration or unusual situations that could lead to a noncompliance are made so proactive steps can be taken to avoid such occurrence. They also verify that no unauthorized or unplanned releases have happened.

Since the results are used to judge the organization's compliance status, the competence of the persons performing the evaluation is critical for reliable results. These persons need to have knowledge and experience commensurate with the scope of the evaluation. Area supervisors can be used to perform facility inspections. Specialized knowledge of legal requirements is needed to perform a regulatory compliance audit, so external consultants may be needed.

The process can be designed to focus on a particular topic, like chemical storage, handling and use, and evaluate compliance across the whole facility, or a multi-media approach can be taken to look at air, water and waste compliance in a specific operation. If external consultants are used, some organizations may find it more practical and cost effective to have them evaluate compliance with regulatory requirements throughout the facility. When organizations decide to use external consultants, they often contract this service on an annual, triennial, or in some cases five year, basis. In the case of triennial or longer schedules, these audits are augmented with routine in-house inspections and other forms of evaluation.

All compliance obligations, whether mandatory or voluntary, need to be evaluated at some frequency. When establishing a schedule, the organization's compliance history, the potential for performance changes and the penalties and ramifications of noncompliance can affect the timing, frequency and depth of the evaluation. New equipment or processes with regulatory requirements may need to be monitored continuously at first to establish confidence that it operates within the regulatory limits, thereby avoiding fines and penalties. Over time, as a history of performance well within the regulatory limit is established, the frequency can be reduced. The regulation itself may dictate a particular time to conduct the evaluation, often aligned with a reporting requirement. An organization may wish to coordinate the timing with assessment of another discipline, for example conducting integrated environmental, safety and health assessments.

The same considerations hold true for voluntary requirements when determining the schedule of their evaluation. An organization may wish to conduct evaluations of customer requirements when they assess conformance to contractual requirements. Depending on the scope, it can be done in combination with the management review process, or separate processes focusing on particular voluntary requirements can be developed and implemented.

Key linkages

- Compliance obligations
- Operational control
- Monitoring and measurement
- Internal audit
- Management review
- Improvement

7.2.1.3 Evaluation methods

Evaluation of compliance can take shape in many formats:

- conduct of facility inspections
- direct observation or interviews
- investigation of employee reports of noncompliance
- review of routine sample analysis or test results
- analysis of verification sampling/testing
- procedure, document and record reviews
- review of regulatory report submittals
- review of product conformance inspections or tests
- conduct of audits
- assessment of activities, products and services to criteria in voluntary agreements and contracts
- review of customer complaints
- review of regulatory inspections and audits

The important factor to keep in mind is that evaluating compliance is an ongoing process. The methods used can be tailored to the organization's size and operations. Also, many of the EMS processes provide information for this evaluation. To avoid duplication of effort, take advantage of the information that is being provided through:

- Monitoring, measurement analysis and evaluation
- Operational controls
- External reporting required by the compliance obligations.

7.2.1.4 Evaluating results

As with any system nonconformity, the cause of a noncompliant situation needs to be identified so that action can be taken to prevent recurrence. If the analysis results indicate a failure to fulfil a legal requirement, the organization needs to determine and implement the actions necessary within its operations to achieve and sustain compliance. This may also entail a review of EMS processes to identify the cause. Potential system causes could include not keeping up with changing requirements, misinterpreting the applicability of a requirement, or not having effective operational controls to ensure compliance, for example.

> A non-compliance is not necessarily elevated to a nonconformity if, for example, it has been identified and corrected by the environmental management system processes.
>
> Compliance-related nonconformities need to be corrected, even if those nonconformities have not resulted in actual non-compliance with legal requirements.

Depending on the applicable requirement, this might also require notification to a regulatory agency or a customer. If the corrective action requires a long lead time to complete, communication and agreement with the regulators on the course of action to bring the operation back into compliance may be required. When an organization enters into such an agreement, it is added to the list of compliance obligations the organization evaluates.

Communicating results of evaluations to top management is required, at a minimum, during the management review. Surely top management would like to be informed without delay if a breach is identified. By evaluating the results from the variety of methods used to self-assess compliance, and tracking the implementation of corrective action, top management can maintain a current and complete picture of its compliance status.

For monitoring purposes and to satisfy regulatory investigation, the organization needs to keep records of compliance evaluation results. The ability to demonstrate consistent record-keeping and retention is important should a breach of regulation occur. It confirms the sense of responsibility within the organization and may, where prosecution for the breach is pursued, help reduce any proposed sanctions against the organization.

NOTE Records are required to be retained as objective evidence of compliance evaluations.

7.2.2 Helpful hints

Since ISO 14001 is one "other requirement", the internal EMS audit performed in accordance with Clause 9.2, Internal audit, satisfies the requirements for evaluating compliance to that obligation (see Section 7.3).

7.2.3 Check conformity

Requirement	Y	N	Follow-up actions
The organization has established, implemented and maintained a process for evaluating compliance with all of its legal requirements, including the frequency for evaluations.			
The organization has established, implemented and maintained a process for evaluating compliance with all of its other requirements, including the frequency for evaluations.			
Compliance is evaluated and action taken if needed if nonconforming/ noncompliant situations occur.			
The organization has a current understanding of its compliance status.			

Requirement	Y	N	Follow-up actions
Records of compliance evaluation results are retained.			
Results of compliance evaluations are communicated during the management review.			

7.3 Internal EMS audit (ISO 14001, Clause 9.2)

ISO 14001 requires the organization to provide evidence that operations and EMS processes are in conformity with the requirements of ISO 14001, and are properly implemented and being maintained. It can achieve this by undertaking a programme of internal audits. This process provides an opportunity for overall assessment of the effectiveness of the EMS so that opportunities for improvement of both the system and environmental performance can be identified and acted upon. Since audit results need to be reported, it also has the benefit of keeping management focused on environmental issues within the organization.

7.3.1 Getting started

7.3.1.1 Taking stock of your current state

Are there other auditing activities conducted in the organization, for example occupational safety and health, quality management or compliance and ethics? There may be a pool of auditors with some of the auditing skills needed for EMS. Consider adopting, or if necessary adapting, their audit procedures.

Quality and safety management systems have similar requirements as EMS, so there may have been some audits conducted that assessed the environmental processes. If not, consider expanding the scope to include environmental processes so that duplication of effort can be reduced.

7.3.1.2 Overview

An audit programme is the arrangements for a set of one or more audits planned for a specific timeframe and directed towards a specific purpose; in the case of ISO 14001, verifying EMS conformity. When planning audits, take into account environmental importance, change and previous audit results. Once the initial comprehensive internal audit is completed, the programme is intended, over time, to focus resources and scope of audits on operations with significant environmental aspects, on changes in the organization that affect the EMS and on EMS elements where internal and external audit results have indicated problems. However, all operations and EMS elements need to receive at least periodic assessment to ensure the system's effectiveness is sustained.

Audit principles

- Integrity and professionalism
- Fair presentation, reporting truthfully and accurately
- Due professional care, applying diligence and judgement
- Confidentiality of information
- Independence, free from bias and conflict of interest ensuring impartiality and objectivity in the audit conclusions
- Evidence-based approach for reaching reliable and reproducible conclusions

The audit programme identifies the responsibilities and requirements for planning and conducting audits, reporting results and retaining associated records. Those that plan the internal audit programme may not necessarily be auditors themselves. However, an understanding of the requirements for conducting audits is essential.

The audit programme stipulates who, what, when and how audits are conducted. Although not specifically required by ISO 14001, the organization may find it of value to document it for training purposes and consistent implementation, particularly where there are multiple EMS auditors.

EMS audit team competency

Technical:

- Environmental aspects and associated impacts, including hazardous substances, pollutants (air, water, soil) and/or natural resources applicable to the organization
- Industry specific process knowledge
- Pollution prevention and control, including emergency protective actions
- Source reduction, waste minimization, reuse/recycling and treatment practices
- Environmental law, reporting and disclosure requirements
- Life cycle perspective, e.g. environmentally conscious design and product stewardship
- Environmental metrics and performance evaluation techniques

General:

- Audit principles, procedures and methods
- Management systems
- Organizational context and associated risks and opportunities
- Applicable legal and contractual requirements

Who — An internal audit programme should be resourced by top management, who should also encourage the recruitment, training and development of a person to manage the audit programme and the internal auditors. The audit manager establishes the audit programme and procedures, monitors implementation progress and reviews the effectiveness of the audit process making improvements as needed. The audit manager recommends the selection of auditors to ensure objectivity and impartiality in the audit process. Independence from the work being audited is also a criterion. Therefore, selection and training of auditors and the identification of audit team leader requires careful consideration, even in a very small company where one auditor may be sufficient.

Training and development is necessary to maintain and develop auditor competence. Competence requirements will depend on the scope of the audit. A single auditor may not possess all of the technical skills themselves, but as a whole, the audit team needs to have the relevant technical skills for the areas and environmental aspects covered by the audit scope. On-the-job training for new auditors can be given during a scheduled audit, provided there is an experienced auditor on the team. Quality system auditors have many of the skills needed to be an EMS auditor. Their skill set can be augmented by taking technical environmental coursework at community colleges or through commercially available training courses.

The use of internal auditors is preferable in terms of cost and knowledge of the operational processes. In some cases, particularly in very small businesses, the relevant skills or manpower may not be available or business priorities may prevent the release of internal auditors from normal duties. In this case, the use of outside auditors with relevant competence is advisable.

What — Organizations must clearly identify the scope and relevant audit criteria when planning audits. This is often referred to as the audit plan.

The overall objective of an internal audit is to determine the organization's conformity to EMS requirements, but there may be cases where a specific audit has a particular intention, such as assessing the environmental performance of a supplier or outsourced process. The audit scope defines the extent and boundaries of an audit — that is, the physical locations, organizational units, activities and processes being audited as well as the dates of the audit.

The audit criteria is derived from ISO 14001 itself and from additional management system requirements the organization self imposes in their policies, processes and work instructions. It is used as the reference against which evidence (e.g. records, statements of fact or other information) collected during the audit is compared to determine whether conformance has been achieved.

Resources also need to be factored into the planning process, including access to information systems, personal protective safety equipment and accommodations for the auditor/audit team to work on site. If facilities are spread out, travel expenditures may be needed, and if external consultants are used, fees will be incurred.

When — Audits can be scheduled so that audit criteria are assessed in increments over a period of time, or can be conducted all at once. A complete audit of all clauses is required for declaring conformity to ISO 14001, but once completed, the audit scope can vary, reflecting priorities and performance trends. Generally speaking, audits should be done at least annually. Documenting audit schedules can demonstrate implementation of the audit programme.

Operations need to be observed, so audits should not be scheduled when activities in the audit scope are not operating. Having said that, it is also important to audit non-standard work operations, such as start-up, maintenance and shutdown, to confirm that environmental requirements are followed. Audits of new operations should be scheduled to confirm application of EMS requirements. Special accommodations to conduct the audit outside of normal business hours may need to be arranged to observe second and third shift operations.

Business activities need to be factored in, so it may not be convenient to schedule audits when a new product is being demonstrated on site for key clients or year-end deadlines are looming.

How — There are three key methods for conducting an audit: document review, interviews and observation of work practices. In general, they follow that order.

The auditor understands how the organization plans to conform to the ISO 14001 requirements and verifies objective evidence of the results that have been achieved by first reviewing the documentation, and then reviewing the procedures describing the process and records of results. This review verifies the documents and records are complete, conform to requirements and are current. This includes verifying control, maintenance and improvement of documented information.

Second, interviews need to be conducted to obtain evidence that work activities are carried out as described in the documentation. Workers can be asked about their access to relevant documents and associated training. Awareness can be audited by asking them how the EMS policy, environmental objectives and EMS requirements apply to their job. Care should be taken to ensure the interviewee is at ease, and that their personal integrity is not being judged. Questions should be open ended and leads them to describe their work activities. Avoid yes/no questions.

Lastly, work activities need to be observed so that evidence of implementation of operational controls is collected. Be aware, under audit conditions, workers may be focused on performing the activity correctly, while under unobserved conditions, may take short cuts. Corroboration through records review, confirms that activities are performed consistently in conformance.

Based on these methods, the auditor evaluates the audit evidence collected in comparison to the audit criteria to determine conformity (i.e. fulfilment of an EMS requirement) and nonconformity (i.e. nonfulfilment of an EMS requirement). Audit findings can also include positive "best practices", potential nonconformities and recommendations for improvement. Audit reports and findings are documented, and reported to management.

The negative findings made during an internal audit trigger initiation of the nonconformity and corrective action process. Revision of the EMS should be made when necessitated. The organization can denote varying levels of nonconformity (i.e. major, minor, observations), to indicate extent or severity. Nonconformities can be analysed for both positive and negative trends. These results, combined with evaluations of compliance, provide the information necessary for top management to draw conclusions on the effectiveness, suitability and adequacy of its EMS, and areas for improvement.

When planning audits, also consider the logistics associated with conduct of the audit, including opening meetings, caucuses among audit team members, and closing meetings to communicate the findings, both positive and negative, made during the audit.

NOTE The audit programme should be documented and kept up-to-date. Records of audit results and programme implementation are required to be retained.

Key linkages

- Competence
- Management review
- Nonconformity and corrective action
- Improvement

7.3.2 Helpful hints

The sections of this guide entitled "Check conformity" can be used as the audit criteria for an EMS audit.

An EMS audit can also be used to conduct a "gap analysis" when an organization first decides to implement ISO 14001. This can serve as an initial review of their processes to identify where additional work needs to be done.

Partner with trade associations, peers or suppliers to sponsor training courses, or conduct internal audits. Costs can be shared, thereby minimized, and has the added benefit that auditors can network and share best practices.

Customers who have imposed environmental requirements or conducted audits at your organization may be willing to share their practices.

Internal EMS audits can be combined with audits of other management systems, such as energy management, occupational safety and health or quality management system audits.

If there are plans to be certified, a good portion of the certification audit will be spent interviewing employees on their awareness. To be sure they are prepared, remember to test the effectiveness of the EMS awareness efforts during the internal audit process.

Documenting the audit dates, scope, criteria, and person(s) conducting the audit, in combination with the audit reported results, fulfils the ISO 14001 requirements for internal audit documentation.

ISO 19011 contains guidance on conducting internal EMS audits.

7.3.3 Check conformity

Requirement	Y	N	Follow-up actions
The organization has established and is implementing and maintaining an internal audit programme that addresses:			
• The frequency of internal audits.			
• The internal audit methods.			
• The responsibilities for managing and conducting internal audits.			
• Planning requirements.			
• Reporting internal and external audit results.			
The frequency and scope of internal audits varies based on importance to EMS effectiveness, changes that can affect it, and takes into account results of previous audits.			
The scope and audit criteria are defined for each internal audit.			
The selection and conduct of auditors reflect impartiality and objectiveness.			
Internal audit results are reported to relevant management.			
Records of audit implementation and their results are retained.			

7.4 Management review (ISO 14001, Clause 9.3)

Periodically, an organization needs to step back and evaluate how it is doing, and if what it is doing brings value to the business. A management review process provides that opportunity for an EMS. It provides a forum for top management to review its environmental performance to see how the results achieved compare to what it set out to achieve. This gives information on the EMS effectiveness. The review also checks whether the processes in place are working as planned in the organization. Do they fit in with its other business processes and make sense for the type of operations it performs and its culture (i.e. the way it accomplishes its work)? This gives information on the EMS suitability. Lastly, the management review verifies that the EMS is meeting the requirements of ISO 14001 and its other compliance obligations. This gives information on the programme's adequacy. Together, these three attributes confirm whether the EMS is working.

A management review is the forum for top management to perform an overall evaluation of the progress the EMS has achieved and to drive future action. This includes mid-course corrections when an environmental objective has not been met. It also drives actions to address changing circumstances that can affect the EMS. This might be related to the changes in the organization's context, interested party requirements, its compliance obligations or significant environmental aspects. The management review also includes actions that might be necessary to sustain the gains realized. Lastly, it includes making decisions on improvement priorities. A management review is the foundation of the strategic planning process for the EMS. Top management is critical to the success of a management review because he/she is the decision maker and opinion leader. It is one of the key responsibilities that demonstrates leadership and commitment to the EMS.

7.4.1 Getting started

7.4.1.1 Taking stock of your current state

What kind of business review meetings does top management currently perform? Does it make sense to add environmental topics to the agenda?

If the organization has other ISO management systems in place, there are management reviews being performed for these other disciplines. Reviewing the EMS at the same time may be efficient.

During which forums do environmental issues get discussed in the organization? Is top management present? Can they be enhanced to address the required topics of a management review and ensure the right people are present?

7.4.1.2 Overview

Information for the management review, gathered from a variety of sources, is clearly of great importance. The information provided enables management to make decisions about EMS suitability, adequacy and effectiveness. The key sources are:

- changes facing the organization that pose risks and opportunities, including internal and external issues associated with the organization's context and changes in interested party requirements;
- results of environmental objectives;
- analysis and evaluation of environmental performance as it relates to environmental aspects;
- internal and external audit results, including evaluation of compliance;
- feedback and concerns relayed during its communications;
- action tracking systems (i.e. corrective actions and previous management review action items).

This information should be summarized for top management so they can see the big picture. Rather than getting bogged down in the details, trends should be presented with an examination of the underlying causes for those trends. The focus is on results achieved, problems encountered, and areas for improvements. What has worked well can provide insight to best practices that can be applied to areas where progress or performance has not met expectations.

The frequency and schedule of management reviews need to be established. Since it is a comprehensive review of the EMS, sufficient time needs to be dedicated to it and the decision makers need to be present. Some organizations find it easier to coordinate shorter meetings at frequent intervals covering a subset of topics at each meeting. This method has its benefits because it can be integrated into existing business review meetings. Others find a holistic review performed less frequently is better because management can focus on environmental issues. ISO 14001 has no requirement for a particular frequency, but many organizations conduct the review annually.

The goal for a management review is to make decisions on a path forward based on conclusions regarding how well the EMS is currently performing, and how anticipated changes can affect its performance in the near term future. Some things to consider may include:

- new environmental objectives, or action needed when current objectives have not been met,

- better ways to conduct EMS processes,
- effectiveness of operational controls, and the need for more, less, or a different approach to controlling environment aspects,
- adequacy of competence, capabilities and resources, and any need for adjustments,
- the effect of internal changes (such as new materials, new product lines or services provided, staffing increases/decreases) on the EMS and if they have been adequately addressed,
- implications of external issues on the organizations' business strategy, and need for changes in environmental policy commitments,
- changes, or new, risks and opportunities that need to be addressed,
- how well the organization fixes identified problems,
- changes in laws or regulations that require a change in approach, update or expansion in the EMS,
- efficiencies that can be realized through new technologies, streamlining EMS processes or better integration into core business processes, and
- priorities for sustaining and improving environmental performance.

Key linkages

- Organizational context
- Interested party requirements
- Environmental aspects
- Compliance obligations
- Risks and opportunities
- Environmental objectives
- Resources
- Communications
- Monitoring, measurement, analysis and evaluation
- Evaluations of compliance
- Internal Audit
- Nonconformity and corrective action
- Improvement

7.4.1.3 Inputs and outputs

ISO 14001 identifies a number of specific inputs and outputs for the management review (see Figure 8). Documentation of the management review results is required. Retain records of the persons in attendance, key discussion points and decisions made. Meeting invitations, agendas, presentation material, meeting minutes and tracking systems typically contain the documented information required.

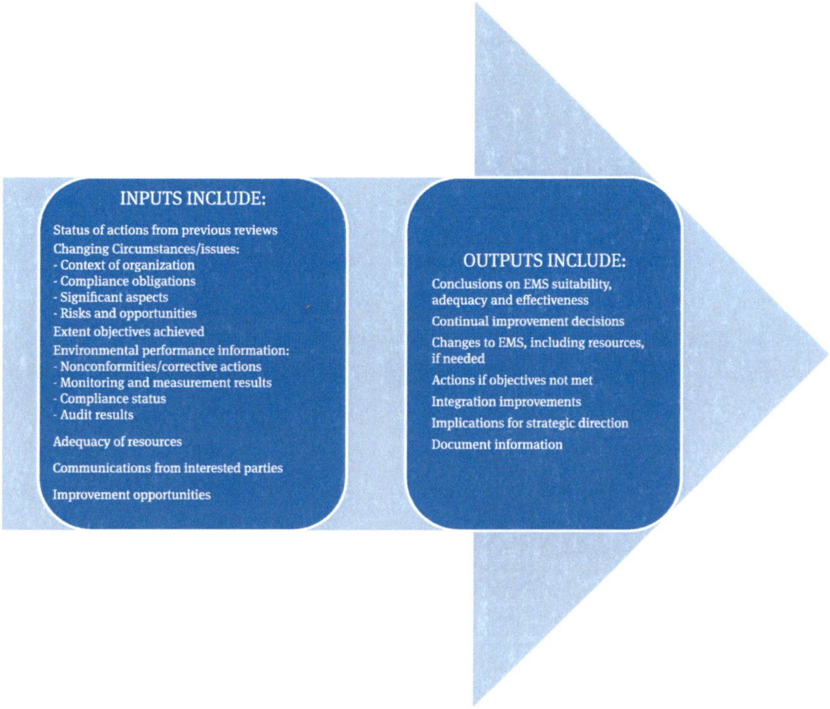

Figure 8 — Management review inputs and outputs

7.4.2 Helpful hints

Appointing a scribe at the onset of the management review meeting can ensure discussion points, decisions and follow-up actions are recorded.

Analysing information needed for a management review can be time consuming. Assign tasks to those responsible for collecting the information giving sufficient notice so that it can be done in time. Preferably they know at the outset the information will be used so that it is routinely generated in a form that can be used directly in the management review presentation material. It is also advantageous to have them present the material to management — they get credit for the good work and see the value of their efforts.

Coordinate the management review with the organization's business planning and budgeting cycles so that decisions made on EMS priorities can be factored in.

Not all topics required in a management review have to be covered in a single meeting. It is usually more effective to integrate these topics into routine management processes, i.e. review key environmental performance indicators, progress on objectives and compliance issues during monthly management meetings.

If the EMS scope was limited initially, the management review can provide the forum for deciding to expand the scope to other parts of the organization.

7.4.3 Check conformity

Requirement	Y	N	Follow-up actions
Top management periodically performs a management review of the EMS.			
A schedule for management reviews is established.			
The information covered during the review includes the required topics.			
The outputs of the review include the applicable topics.			
Records of the management review are generated and retained.			

8 Improving the EMS and environmental performance

Improving the EMS and environmental performance are critical components of the Act phase of the Plan-Do-Check-Act model. ISO 14001, identifies three clauses to assist the organization in these efforts:

- improvement to achieve the intended outcomes (10.1),
- nonconformity and corrective action (10.2),
- continual improvement of the EMS (10.3).

For ease in understanding, this guide explains nonconformity and corrective action first, and explains the remaining two concepts together.

8.1 Nonconformity and corrective action (ISO 14001, Clause 10.2)

Despite the best efforts, inevitably the EMS processes will not be implemented as planned or the results will not achieve what was desired. Equipment malfunctions; a process or procedure is not followed; a supplier sends the wrong material; a wastewater discharge exceeds the regulatory limit; an audit reveals an EMS requirement is not implemented. No EMS is perfect, so it is important to implement a systematic method for identifying the cause of problems that occur and taking action to correct them and prevent their recurrence.

8.1.1 Getting started

8.1.1.1 Taking stock of your current state

Organizations with a quality management system already have a process for detecting and correcting problems. The same process can be used if it works effectively and expanded to address environmental problems.

What problem-solving processes exist in the organization? Implementation of lean, six sigma, TQM or other improvement methodologies often use multiple tools, like the "5 why" process, fishbone diagrams or failure mode and effects analysis to solve problems.

8.1.1.2 Overview

A nonconformity occurs whenever an EMS requirement is not fulfiled or when environmental performance requirements are not met. The organization is required to identify and correct nonconformities and first take action to mitigate any environmental impact, then analyse the cause(s) so that the necessary and sufficient actions can be taken to prevent them from occurring again in the future.

Information on deficiencies can be received from different sources such as emergency situations, monitoring data, audit reports and evaluations of compliance. Information from employees (either directly or through suggestion boxes or "near miss" reporting processes, for example) are other channels where potential problems can be identified. It is important to establish communication channels so employees can talk openly about situations that could lead to breaches in compliance or polluting incidents without fear of sanctions.

If the problem involves a polluting event, the organization must promptly address the immediate situation to correct it and mitigate any environmental impact. If, for example, a spill occurs during a refuelling operation, immediate action to stop and contain the petroleum spill, clean it up and, if necessary, notify the authorities.

Once the environmental impact has been mitigated, the organization can move on to analysing the nonconformity. Initially it may appear there is an obvious cause. For example, if the refuelling process was inadequate, or an employee was not competent in the controls needed during refuelling. It is important to look beyond the immediate reason for the nonconformity to determine why it occurred. Involving those directly responsible for the operation where the problem occurred can help in identifying the failures that caused the nonconformity. Was there an equipment malfunction? What conditions provoked the error, e.g. multitasking, fatigue, visual displays. Do procedures cover normal and abnormal situations? Was training material lacking? There may be multiple causes that will need to be addressed in order to prevent the problem from recurring.

Corrective actions address the failure in the EMS element that caused the problem. Where a performance target was the source of the nonconformity, consider adjustments to the schedule, allocated resources or targeted result. Be sure the responsibility and schedule for correcting problems are assigned and that the actions are implemented in a timely fashion. Process changes need to be captured in EMS documentation and formally communicated to the affected employees. Where substantial changes are made, it may be necessary to hold a more formal briefing in which the nature of, and justification for, the changes can be fully discussed.

Preventing problems before they occur is one method of improvement. When a nonconformity occurs in one location or part of the organization, it is worthwhile to look for situations in other locations or parts with similar activities and implement actions to prevent the problem from occurring there.

A periodic review contributes to the effectiveness of the EMS. By reviewing the results of corrective actions, the organization can determine if the change improved performance and was successful in avoiding additional breaches or failures.

The analysis, timing and documented information associated with nonconformity and corrective action should be proportional to the severity of the problem and the associated environmental impacts that could arise. The allocation of resources to manage an EMS is always uppermost in the mind of top management. Resources should be fairly allocated to those nonconformities presenting the highest potential adverse effects.

NOTE Records are required to be retained as objective evidence of the type of nonconformity, action taken to correct and prevent its recurrence, and effectiveness of the results (see Figures 9 and 10).

Key linkages

- Emergency preparedness and response
- Monitoring, measurement, analysis and evaluation
- Evaluation of compliance
- Internal audit
- Management review

Communicating

- Corrective actions that result in changes to operations or EMS processes need to be communicated to those responsible for implementing the change.
- A summary of the types of nonconformities that occurred, and trends in how effective the corrective action process is in dealing with them should be presented for evaluation during the management review.

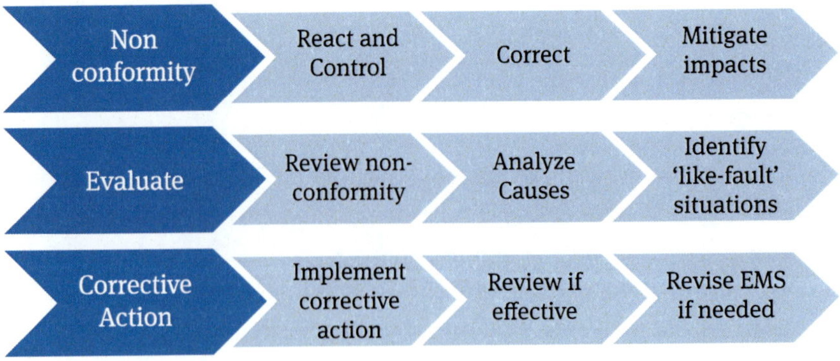

Figure 9 — Overview of nonconformity and corrective action process

CORRECTIVE ACTION REQUEST FORM			
Section 1		**General information**	
CAR #:		*Date:*	
Area/department:		*Area manager:*	
Section 2		**Nonconformity statement**	
Major ☐ Minor ☐		Reference to requirement:	
Signature:		Date:	
Section 3		**Corrective action**	
Immediate action		*Completion date:*	
Causal analysis and like-fault results			
Corrective action(s)		*Scheduled completion date:*	
Results of effectiveness review		*Date:*	
Mgmt. signature:		Date:	

Figure 10 — Example of corrective action request

8.1.2 Helpful hints

If the nonconformity relates to a failure in operational controls, consider the hierarchy of controls when developing corrective action, eliminating the environmental aspect that was the source or implementing an engineered solution before considering administrative controls.

Identify trends in nonconformities so that future problems can be prevented.

Share information on nonconformities and lessons learned, during staff meetings and by incorporating it in training materials.

8.1.3 Check conformity

Requirement	Y	N	Follow-up actions
Take action to control and correct nonconformities.			
Mitigate adverse environmental impacts and conduct follow-up actions if necessary.			
Review nonconformity to determine the causes and if similar nonconformities exist, or can exist, elsewhere.			
Evaluate the need for action to prevent recurrence.			
Evaluate the need for action to prevent occurrence elsewhere.			
Implement corrective actions.			
Review effectiveness of corrective actions.			
Update EMS to reflect any changes made.			
Retain records of: • Type of nonconformities • Corrective actions taken • The results of corrective actions.			

8.2 Opportunities for improvement (ISO 14001, Clauses 10.1 and 10.3)

Improvement is a critical and integral part of the EMS. The focus of improvement efforts is toward achieving the EMS intended outcomes, which include enhancing environmental performance, fulfilling its compliance obligations and achieving its environmental objectives.

The EMS framework is designed to provide the organization with the information and processes needed to identify areas of improvement, prioritize them and implement them. As such, by complying with the ISO 14001 requirements in Clauses 9 and 10.2, actions to improve will be taken and improvements made without necessarily taking any additional action; the EMS itself drives improvement.

8.2.1 Getting started

8.2.1.1 Taking stock of your current state

The EMS processes associated with the "Check" step of the Plan-Do-Check-Act model provide information for identifying areas for improvement. The EMS processes associated with the "Act" step and "Plan" step provide processes for implementing the identified improvements.

8.2.1.2 Overview

The two clauses of ISO 14001 specifically addressing improvement are Clauses 10.1 and 10.3. They require the organization to plan for and take action to implement improvement. The first requires improvements to achieve EMS intended outcomes; the second addresses improvement to the EMS itself.

The form of improvement can be as simple as fixing a single identified problem, perhaps in response to an identified nonconformity. Alternatively, it can entail reimagining and redesigning products and services from an environmental perspective to create something totally new that creates a competitive advantage for organization. For example, Toyota illustrated this with the introduction of a hybrid car.

Or it can involve improvements performed in series over time focused on the intended outcomes. The organization may set an environmental objective and action plan that entails installation of new control technology to control air emissions. Later, after the equipment is installed and initial results gathered, operational controls settings may be optimized to ensure it consistently meets air emission permit limits. Later still, the organization may for example, identify and implement reduction, reuse and recycling options for the various waste streams.

Recurring improvement activities, such as these, is referred to as "continual improvement". Improvements do not have to take place in all areas simultaneously, nor do they have to be conducted without interruption. The rate and extent to which an organization takes action to improve will be based on their business priorities and resource capabilities.

In addition to making improvements to achieve intended outcomes, ISO 14001 requires an organization to continually improve the suitability, adequacy and effectiveness of the EMS itself. This entails improving the suitability of the individual EMS elements over time so that they are adapted to fit the needs of the organization. It might also address improvements in order to meet current and new requirements. Lastly, it is about improving the system elements that drive results. The goal of these improvements, is on enhancing environmental performance, therefore focusing specifically on those elements that will contribute to improvement in the management of environmental aspects.

The EMS elements that can be used to mitigate environmental impacts include environmental objectives, operational control, and emergency preparedness and response. Planning and implementing improvements to these processes can lead to improvement in environmental performance.

Improving other EMS elements may have an indirect but nevertheless beneficial effect on environmental performance, so they should also be explored.

- Training employees on material reuse and recycling can reduce the generation of waste.

- Enhancing the root cause analysis used to investigate nonconformities can not only prevent recurrence of the problem, but can prevent similar problems from occurring elsewhere.
- Standard operating procedures that are streamlined and transferred into visual instructions, make it easier for employees to follow environmental requirements.
- Training materials revised to incorporate lessons learned from near-miss incidents and emergency situations can prevent future problems from occurring.

As improvement activities progress through implementation, not only will the EMS processes be improved, but so will outcomes improve.

Key linkages

- Environmental aspects
- Environmental objectives
- Compliance obligations
- Monitoring, measurement, analysis and evaluation
- Evaluation of compliance
- Internal Audit
- Management review

8.2.2 Helpful hints

Responsibility for improvement initiatives can be assigned and deployed through the personnel performance goals and appraisal process.

Techniques for identifying and implementing process improvement can include the six sigma tools such as Lean, Value stream mapping and Kaizen.

Benchmarking with "best in class" companies can provide information on continual improvement.

8.2.3 Check conformity

Requirement	Y	N	Follow-up actions
Opportunities for improvement are identified through the following EMS processes:			
• Monitoring, measurement, analysis and evaluation			
• Internal Audit			
• Management review			
Improvements needed to achieve the EMS intended outcomes are prioritized and implemented.			
The organization continually improves the suitability, adequacy and effectiveness of its EMS in order to improve the management of its significant aspects.			

9 Integrating other management systems

Management system integration is becoming increasingly complex with the growing number of management system standards — food safety, financial management, occupational health and safety, information security, as well as business continuity, in addition to quality and environmental management. Integrated or harmonized systems represent a commercially and financially sensible approach to the implementation of a number of discipline-specific systems. To address the needs of organizations implementing multiple management systems, ISO developed a unified structure and set of core requirements and common terms for use in all management systems. Management systems developed or revised after 2012 are required to conform to these requirements. Both ISO 14001:2015 and ISO 9001:2015 conform to these requirements.

Organizational benefits of harmonized and integrated management systems include:

- common processes,
- fewer documents to manage,
- consistent and integrated training for employees,
- reduction in audit costs,
- streamlined and consistent implementation,
- competitive advantage.

A coherent management system approach makes it easier for top management to plan, implement and control operations, and provides methods for checking results to achieve continual improvement. The Plan-Do-Check-Act model is the basis for such a coherent system (see Section 1.2). Integrating environmental management, quality, food safety and other disciplines may, even in a small company, provide the basis for a comprehensive business system. A coherent approach can raise the prominence of the discipline, in this case environmental management, as an important organizational issue. When planning and implementing an integrated management system, it is important to acknowledge that each discipline has a unique purpose. Its requirements need to be applied correctly and comprehensively to the relevant activities, products and services of the organization. Each organization can decide for itself the extent to which it will integrate their management processes. Often, integration increases as the system matures and evolves.

An organization can seek third party certification to one or more of the ISO management system standards; however, just because it is an integrated system does not mean it must have all disciplines certified. If an organization does seek certification of multiple disciplines, the audits conducted by a third party can be performed simultaneously on an integrated management system, saving time and costs.

Table 16 outlines management system processes that have the most commonality, which then lends itself to integration, versus discipline specific processes, which are unique to an EMS. It is important to note that there are processes which can be similar but must be applied to each discipline, taking into account the unique considerations of that discipline. For example, the process used to identify interested parties may be common, but the interested parties relevant to an EMS will be different from those that are relevant to a QMS. A common process for setting and monitoring objectives can be used but the subject matter of environmental objectives will obviously be different.

Table 16 — Common versus Discipline-specific management system processes

Common processes	Similar processes	Discipline-specific processes
Organizational context	Interested parties	Environmental aspects
Leadership	Policy	Compliance obligations
Competence	Roles, responsibilities and authorities	Operational control
Resources	Risks and opportunities	Emergency preparedness and response
Documented information	Objectives and planning action	
Calibration	Awareness	
Internal audit	Communications	
Nonconformity and corrective action	Monitoring, measurement, analysis and evaluation	
Continual improvement	Management review	

In the planning phase of implementation, organizations should consider the following steps.

- The context of the organization can serve a multi-disciplinary system.
- The scope of the system: wherever possible, each discipline should be implemented within the same boundaries and functional areas.
- Policy: organizations should ensure that their policies are consistent in language, form and approval authority.
- Compliance with environmental, financial, safety and other regulations are paramount considerations for management, and treated equally across all disciplines.
- Risks and opportunities should be evaluated holistically, across multiple disciplines, to ensure balanced priorities.
- Objectives and action plans: the establishment objectives for all disciplines should undergo a common approval process.

In the doing (implementation) phase, organizations should consider the following steps.

- Allocation of roles, responsibilities and resources: there can be efficiencies when assigning a functional responsibility (i.e. auditing, training, etc.) for more than one discipline to a single individual. If this is the case, care is needed to ensure adequate competence across all disciplines and to balance resource needs for each discipline. Competence and awareness needs are likely to differ significantly between disciplines, but the processes used to determine needs, assess competence and retain records should be managed in a coherent way.
- Communication: with a multi-disciplinary system in operation, organizations can coordinate the flow of information and ensure consistency and appropriateness for the target audience.
- Documentation: consistent processes, style and language can be used to ensure coherent implementation across disciplines.
- Operational controls: day-to-day work instructions need to be clear and comprehensive to reflect a multidisciplinary approach and enable workers to achieve the activity's specific objectives.

In the checking phase of implementation, the internal audit becomes very important in ensuring a technically balanced view of the effectiveness of each discipline's management system elements. Therefore, the auditor/audit team competence in the various technical disciplines is important.

With the implementation of multiple disciplines, the act phase of implementation is important in ensuring balanced priorities and adequate resources when setting improvement priorities for each discipline.

10 Demonstrating conformity

10.1 Conformity assessment options

An organization can implement ISO 14001 without seeking certification. The best reason for implementing ISO 14001 is to improve the efficiency and effectiveness of the organization's operations. Certification of the management system is not an ISO 14001 requirement.

Implementation and development of an EMS in accordance with ISO 14001 can demonstrate an organization's responsible approach to environmental protection, and its commitment to compliance with environmental regulations. The choice of method to demonstrate such responsibility and commitment will depend on the organization's relationship with its customers and other stakeholders.

Organizations are offered several choices in demonstrating conformity with ISO 14001:

1) making a self-determination and self-declaration, or
2) seeking confirmation of its conformance by parties having an interest in the organization, such as customers, or
3) seeking confirmation of its self-declaration by a party external to the organization, or
4) seeking certification of its environmental management system by an external organization.

Making a self-determination and self-declaration is a simple approach and depends largely on the organization having a robust internal audit process in place. The organization assesses itself and states its conformance with the requirements of ISO 14001 using its own resources and then publishes its findings.

Confirmation of conformity is frequently achieved through second party audits. This offers an independent view of the suitability and effectiveness of the EMS. The relationship between the two parties, and self-interest of the party performing the audit may influence the scope, criteria and objectiveness of the results.

Confirmation of a self-declaration by an independent third-party is another, possibly more credible service offered by an environmental consultant, or by a certification body. While it might be less costly, this form of validation lacks the rigour of ongoing audits that ensure the EMS is maintained and improved.

The organization can make a declaration of conformity using any of the three options described above. ISO/IEC 17050-1 sets out the minimum requirements for such a declaration. Figure 11 shows a typical form for declaration of conformity.

Certification is often seen as a costly burden on the organization, particularly for smaller organizations. It is important, therefore, that the certification body provides an audit programme covering the full three year certification cycle with all audit costs, including audit day rate, fees, and travel estimates, clearly identified.

Declaration of conformity		
1) Identification number:		
2) Issuer's name:		
3) Issuer's address:		
4) Object of the declaration:		
The object of the declaration described above is in conformity with the requirements of the following documents:		
Document number	Title	Edition/date of issue
5)		
Additional Information:		
6)		
Signed for and on behalf of:		
(Place of issue)		
7)		
(Name, function)	(Signature)	(Date of issue)

Figure 11 — Example of form for declaration of conformity

10.2 Value of certification

ISO 14001 is recognized by customers, regulators, and others as a means of demonstrating a responsible approach to protecting the environment and minimizing adverse impacts. The global recognition of ISO 14001 makes it ideal for companies involved in export activities. Many organizations who wish to avoid the costly process of undertaking their own procurement audits therefore seek suppliers who are certified to ISO 14001. Certification is mandatory for suppliers in some industries; for other industries, it could enable a supplier to obtain a preferred status, thereby providing a competitive advantage.

Financing charges and insurance costs may be treated more favourably if the company can demonstrate that it is not at risk of breaches in environmental regulation, with the threat of consequential penalties. Regulators may reduce inspection priority or other incentives for facilities with ISO 14001 certification.

Ongoing audits throughout the certification cycle keep attention and focus on the EMS within the organization, ensuring that the system is maintained and improved. Certification audit results give valuable input to areas where the organization should focus their improvement to achieve both environmental and business benefits.

While certification is not a guarantee, adherence to the system, watchful monitoring and performance reviews substantially reduce risks. Consequently, it can help build confidence in the organization.

10.3 What to expect when becoming certified

There are several steps to becoming ISO 14001 certified and to maintaining this status once achieved.

An initial on site audit — often called a Stage 1 audit — is carried out by the certification body. It involves a review of the management system design, including select documents, to ensure that the organization has implemented the key requirements of ISO 14001. The auditor identifies any gaps (nonconformities) found in the system for corrective action by the organization.

After satisfactory completion of the Stage 1 audit and correction of any major nonconformities, the certification body auditors undertake the Stage 2 audit. This involves a site visit to audit all ISO 14001 requirements and assess appropriateness and degree of EMS implementation. On completion, the auditor once again identifies any nonconformities found in the system for corrective action by the organization. Major nonconformities will need to be corrected and verified before the certification body can make a positive recommendation for certification. The final decision is made by the certification body's management oversight board, and once approved, a certificate with instructions on using the accreditation mark are provided to the organization.

Certification covers a cycle of activities over a period of three years. During this time, periodic surveillance audits (once or twice a year) are carried out by the certification body to ensure that the EMS remains suitable, is maintained and improves. In the surveillance stages of the certification audit cycle, auditors will also wish to assure themselves that previous nonconformities have been addressed.

It is important that the certification body provides an assessment and audit plan for each audit, setting out the date and time of visits, EMS elements to be audited, organizational activities to be observed and interviews to be conducted, etc. This should enable the organization to be properly prepared, with guides and safety information provided if necessary. The organization will need to be prepared to accommodate the needs of the audit team. The audit team leader will require rooms, equipment and internet access to conduct opening and closing meetings, interview staff, hold confidential progress meetings with other team members and prepare reports. If EMS documents and records are stored on a secure intranet, a protocol for accessing the relevant information without compromising operations, commercial confidentiality or data is necessary.

Certification audits typically take the form of reviews of management system documentation, combined with observation of EMS-related work processes and interviews of persons at all levels to identify the extent to which employees understand their obligations within the EMS. They are not intended to be stressful experiences for employees and should be conducted in a constructive and educational context. Typically, he/she will ask open questions, requiring more explanation than a "yes" or "no" answer and will seek to corroborate answers with information documented in the system. It is prudent to explain the auditor's role to employees in anticipation of the certification audit, together with encouragement to answer all questions in a concise, honest and balanced manner.

Audit findings, including nonconformities, are identified by the auditor and shared during the daily debriefs. It is important to discuss these findings so that both parties can clarify the issue and improve understanding. There may be cases where it is appropriate to challenge a finding; perhaps there is additional objective evidence that was not initially provided to the auditor. In the end, however, it is the auditor's decision that stands. The organization has an appeal process available through the Certification Body should it need one.

A formal audit report will be provided following each onsite visit. This will present the audit conclusions, request corrective actions for nonconformities in accordance with the timetable agreed at the closing audit meeting. The organizational representative signs the audit report submitted at the closing meeting, acknowledging receipt of the findings.

The organization may need to plan for a special audit in the event a significant nonconformity in the system prevents the certificate award, a complaint has been lodged, or a major change in the organization's structure and processes.

Requirements for certification bodies conducting third-party certification audits of management systems can be found in ISO/IEC 17021-1, requirements for competence of certification bodies performing EMS audits can be found in ISO/IEC 17021-2, and guidelines specifically on duration of third-party EMS certification audits can be found in ISO/IEC 17023.

10.4 Choosing a certification body

Choosing a suitable certification body to undertake the certification audit is a critical part of the certification process. In making the selection, an organization may wish to consider:

- If the certification body is accredited for environmental management system audits by an accreditation body.
- The auditors are experienced in certifying organizations in a similar industry sector.
- Overall costs for the full three year certification cycle (Stages 1 and 2, surveillance and reassessment audits, plus travel estimates). The lowest price, however, does not always represent the best value for money.

- Client references and referrals from peers and industry groups.
- The certification bodies process for handling and reporting regulatory noncompliance, including how identified noncompliant situations affect their decision-making regarding granting certification.

Certification/registration bodies are not consultants and are not, therefore, advisors on the implementation and management of the EMS. However, responsible certification/registration bodies do seek to try and provide added value to their clients by clarifying ISO 14001 requirements when needed and constructively challenging the system.

10.5 Certification audit and legal compliance

The certification auditor does not act on behalf of enforcement agencies and is not responsible for his/her client's compliance with legislation. In terms of confidentiality, the auditor's primary responsibility should be to the client. However, there are circumstances where breaches of legislation and the potential for pollution may be so severe that the professional reputation of the certification body could be damaged by association. In these circumstances, the auditor will notify the client and follow the agreed upon process for handling and reporting regulatory noncompliance.

It is a requirement that the audit team as a whole be competent to audit environmental legislation for which the organization is liable. The system documentation should identify regulatory requirements with which operations should comply. During the audit, the certification body will assure itself that:

- a methodology is in place to identify, maintain and update relevant environmental legislation
- the regulatory requirements are used to design the EMS, operational controls, and monitoring processes
- the organization's compliance status with respect to applicable regulatory requirements can be demonstrated.

If audit evidence suggests that there is failure in any one of these requirements, it will be identified as a nonconformity. This does not mean that the organization automatically fails the certification audit. The auditors will verify that the EMS has processes to handle noncompliant situations (i.e. nonconformity, corrective action, communication, etc.), and that overall the noncompliance does not rise to the level of a major nonconformance in the system. It is imperative, therefore, that the organization respond promptly to correct any such circumstances following any such protocols, otherwise a report to the appropriate officials may be necessary.

11 Case study — Example of ISO 14001:2015 implementation in a small company

Objective

The case study in this annex shows how a small fictitious manufacturing company has undertaken an EMS implementation project. The intention is to provide some examples of how EMS elements can be applied to a mock company. The examples provided are not intended to be applicable to similar organizations, nor are the examples given intended to be exhaustive.

11.1 Description of the organization

SB is a metal fabrication plant located in the outskirts of a large city. It was founded in 1935 close to a channel, a tributary of the river that crosses the city. John (the owner/manager) inherited the company from his father in 1969.

In 1935, the plant was located in a rural area that was sparsely populated. In the last 20-30 years, urban sprawl engulfed the area. The plant is now surrounded by densely populated suburbs inhabited by middle class families who have escaped from the downtown smog and traffic. There is also a large community square nearby (50 metres away from the plant fence line) that is being used by the residents on a daily basis.

SB manufactures different types of small components for the automobile industry. Its principal client is a large car manufacturing company that is also located in the same area. Production starts at 6:00 am and ends at 2:00 pm. The first batches of product are ready for distribution at 6:00 am. The component parts are packaged in plastic bags then packed in 10-50 kg wooden boxes. The products are loaded from 6:00 am onwards, using nine vans contracted from a logistics company. This is to comply with their customer requirement for "just in time" delivery of spare parts.

The raw material used by SB is in the form of thin plates and bars made from different types of steel. The raw material is cut to the desired dimensions using two lathes, some are thermally treated in three small furnaces fired by natural gas. The lathes and furnaces are controlled semi-automatically. Raw materials (steel, lubricant oils, chemicals, packaging material) are stored in the basement. The manufacturing processes are performed on the ground floor and several administrative offices are housed on the first floor. There is no loading dock so the vans are parked on the street while the products are loaded.

Within the last 20 years, a primary school (with 500 pupils) opened across the street from the plant. More recently, construction of a small hospital began on the corner of the community square.

John manages SB, with the help of three supervisors (production/maintenance, logistics/distribution and purchasing/raw materials storage). The production line has 17 workers dedicated to the manufacturing processes, while three more staff members take care of infrastructure maintenance.

John will retire in the next few years, and plans to hand over the company to his two children: Hillary (an industrial engineer), who is currently employed by the parks division of the local city council responsible for maintaining the public gardens and community square; and Bill who will graduate from college in a few months with a degree in economics.

Hillary and Bill are planning to expand the plant and convinced John to buy a piece of land close to SB. A new building and machinery is being built so they can start production of a new line of products, plastic vehicle components, for its principal client. They also plan to fully automate the cutting machines and lathes.

11.2 SB and the environment/setting EMS expectations

John was always concerned about the environmental impacts caused by the SB operations and took his legal obligations very seriously. He explained to Hillary, "We have the legal authorisation to operate the company which includes an environmental certificate demonstrating that SB is in full compliance with the environmental regulations". He thought: "Why I am going to do more? My contribution to the deterioration of the environment is minimal".

However, during a meeting with Bill and Hillary, attention focused on a series of spills that had occurred in the chemical storage area in the basement. Although the spills represented a modest cost to SB due to the loss of raw materials, the clean-up operation added substantially to the cost of waste disposal. Moreover, it was acknowledged that these incidents were inconsistent with the company's public commitment to prevent pollution on its property. Advice from the regional Association of Metallurgical Organizations also indicated that there was a legal requirement to report spills to the environmental regulatory agency, which had not been done.

John recognised this was a serious issue and decided to initiate an environmental management project to address the problem. An environmental consultant in the Association suggested two actions:

- **a)** an immediate solution for the current spillages that would prevent them from occurring, and
- **b)** a longer term solution that incorporates a more systematic and proactive approach to protect the environment by identifying and complying with all legal requirements.

By taking this two pronged approach, SB would ensure that it could sustain its commitment to prevent pollution.

Following the regional Association recommendation, John decided to implement an environmental management system (EMS) as described in ISO 14001:2015. The SB management team realised that the intended outcomes of the EMS (enhancing environmental performance, fulfilling compliance obligations and achieving environmental objectives) would fully address their concerns and expectations. Bill and Hillary attended a course organised by the Association on ISO 14001 and helped John develop a plan to meet the requirements in a way that made sense for a small company such as SB.

11.3 Context, interested parties and risks and opportunities

There were a number of factors that SB needed to consider when planning its EMS. First and foremost was how the spills from SB´s operations could affect the environment; and secondarily, how the local environmental conditions could affect its operations, in particular flooding by the nearby channel. There were other considerations, such as its internal expansion plans and noise complaints expressed by the growing neighbourhood. ISO 14001 required SB to evaluate these internal and external factors to understand its "context" and determine if there were any risks and opportunities (Clause 6.1) that could either hinder or help them in achieving their EMS intended outcomes. By doing so, SB could be sure to take action to address them within the EMS in a prioritised manner.

John, Hillary and Bill, SB's "top management", met several times to identify and evaluate the potential risks and opportunities associated with 1) the current issues related to the context within which SB operates (Clause 4.1), and 2) the relevant needs and expectations of internal and external interested parties (Clause 4.2). The priority level assigned to each issue or requirement was decided (through consensus) by the three members of SB's top management based on the degree it could affect SB's ability to achieve the intended outcomes of its EMS, prevent adverse effects, or lead to an improvement. They also decided to document their analysis as shown in Table 17 even though ISO 14001 requires documented information of the risks and opportunities only.

Further, an in-depth analysis of risks and opportunities associated with the environmental aspects of SB's activities, products and services (Clause 6.1.2) was conducted. Three potential emergency situations were identified, that were included in the evaluation:

- flooding of the channel,
- spills of stored chemicals/lubricants,
- fire.

The criteria for determining if the environmental aspect was significant was based on the potential severity of the environmental impact. A portion of the evaluation is extracted and presented in Table 18 for illustrative purposes.

The construction of the new building was contracted to a local engineering firm. All environmental regulations and permits were taken into account in the contract specifications. Compliance will be checked under John´s supervision. Further, a preliminary review of the future production processes was conducted under the newly implemented EMS framework and SB´s environmental policy. The new plastic molding process will introduce environmental aspects and may require additional operational controls. John and Hillary scheduled a meeting with the Association to seek information about the environmental impacts of plastic molding.

Table 17 — Analysis of SB organizational context, interested parties and risks and opportunities

Organizational context issue		Risks/opportunities	Priority	Action
External context issues				
Environmental conditions affecting SB	Flood of the channel close to the plant (twice per year)	Flood in the basement (raw materials storage)	high	Move storage to the ground floor of the new building
Other conditions	Technology	Advancements in production technology	medium	Plan to update machinery as funding allows
	Environmental regulations	Not aware of legal requirements	high	Seek assistance from Association of Metallurgic Organizations
	Market competition	Competitors having a “greener” reputation within the community	medium	• Benchmark competitors • Improve communications with the community
	Social / reputation/ public image	Parking and loading of vans in the street creates nuisance to traffic and school	high	When designing the new building, allow space for parking inside SB´s premises
Organizational context issue		**Risks/opportunities**	**Priority**	**Action**
Internal context issues				
Organization		John handing over company to Bill and Hillary	low	No action. John´s retirement will be gradual
Organizational knowledge		Loss of knowledge due to the retirement of 2 of the 3 supervisors	high	Find replacement and ensure an overlap of 6 months minimum
Strategic direction		SB´s capabilities exceeded with new expansion	medium	Carefully plan the infrastructure and personnel needed for the expansion
Activities, products and services		Several environmental impacts	high	Establish operational controls (see Table 19 below)

Interested party	Need/expectation	Priority	Action
Interested party needs and expectations			
Employees	Work conditions	low	No action
Suppliers	Long-term planning	low	No action
Neighbours	Noise levels are too loud when operations start-up	high	Establish environmental objective to reduce noise level
Regulators	Fines if regulatory reports are not filed on time	high	Establish a process to ensure reports are filed on time
Customers	Products and services expected to be as "green" as possible	medium	Introduce the "life cycle perspective" in product design
Environmental charter of the Association of Metallurgic Organizations	Opportunity to improve SB´s reputation in the community	high	Analyse requirements

Table 18 — SB environmental aspects and impacts

		Environmental aspect	Environmental impact	Significant? (based on severity of impact)
Processes	Materials storage in the basement	Spills of chemicals	Underground water contamination	yes
	Production	Potential fire	Water and air contamination	yes
		Energy consumption	Depletion of natural resources	yes
		Discharge of water to the channel	Channel contamination	yes
		Emission of furnace vapours	Air contamination Elevated noise levels	yes
		Cooling water consumption	Depletion of water resource	no
	Infrastructure	Air conditioning	Energy consumption	no
		Use of solar cells for heating water and central heating	Energy consumption	no
	Administrative	Use of copier	Disposal of hazardous materials	yes
		Use of paper	Depletion of forests	no
Products	Packing and packaging material (use and disposal stages)	Disposition of paper, plastics and wood by the customer	Depletion of natural resources Land contamination	yes
Services	Distribution (outsourced process)	Vans consumption of fuel	Depletion of natural resources	no
			Air contamination	yes

With the assistance of a lawyer working on behalf of the Association, the legal requirements (Clause 6.1.3) were identified for SB. The following list of legal requirements were noted:

- environmental permit to operate (local authority)
- periodic reports on environmental performance of the operations (air emission, noise, water consumption, discharge of liquids to the channel, hazardous waste generation)
- hazardous waste disposal requirements
- local town restrictions related to vehicle use in the community (vans, maintenance, drivers, etc.).

Other compliance obligations (i.e. the voluntary ones that SB chooses to comply with) were identified during the risks and opportunities identification process, and included:

- Environmental charter of the Association of Metallurgic Organizations
- ISO 14001
- Resolution of noise complaints from neighbours.

Hillary agreed to document and review the lists of environmental aspects/impacts and compliance obligations to be sure they remain up to date, and revise them when necessary as required by ISO 14001. John and Bill agreed to review any changes during the management review (Clause 9.3).

SB's top management decided to address all legal requirements during the first year of the EMS implementation project, and address the voluntary requirements starting in the second year. They also agreed the risks and opportunities rated "high" and the environmental aspects rated as significant would need to be addressed in the first year, and would re-evaluate the priority of the remaining items the following year.

Environmental objectives and the plans to achieve them (Clause 6.2) were established to address improvement opportunities, and are described in the Section 11.9, "Driving improvement". Since SB's standard practices already addressed its significant aspects and legal requirements, it made sense to incorporate any additional controls needed into the routine operation of SB´s business processes (Clause 8.1). The approach taken by SB is described in the clause "Operations/integration with business processes" below. Progress on objectives will be reviewed during the management review meetings (Clause 9.3).

11.4 Operations/integration with business processes

The actions to augment SB's business processes with additional operational controls are shown in Table 19 below.

Table 19 — SB operational controls

Environmental aspect	Business processes affected	Operating criteria
Spills of chemicals	Logistics/ human resources	• Train supervisor and personnel on handling chemicals • Make available safety data sheets for all chemicals • Build some type of containment to prevent spills to reach the drainage (the storage area will be moved to the ground floor of the new building in 4-6 months' time)
Potential fire	Production/ maintenance/ human resources	• Train supervisor and personnel on fire prevention • Establish a maintenance plan for the electrical and natural gas installation • Identify and store separately all combustible materials
Energy consumption	Production/ maintenance	• Control furnace efficiency by monitor-ing the O_2 content of exit gases • Establish a maintenance plan for furnaces and insulations
Discharge of water to the channel	Production	• No proactive actions needed; current levels of contaminants demonstrate compliance with regulations. Keep monitoring periodically
Emission of vapours from furnaces	Production	• No proactive actions needed; current levels of contaminants demonstrate compliance with regulations. Keep monitoring periodically
Disposition of hazardous materials	Logistics	• Maintain the current process of hazardous materials handling

Environmental aspect	Business processes affected	Operating criteria
Disposition of paper, plastics and wood by the customer	Sales	• Continue using reusable packing and packaging materials • Keep customers informed on possible alternatives for reusing these materials • Include simple instructions to customers on how to reuse or dispose of the wooden boxes and packing material
Vans consumption of fuel	Logistics (outsourced process)	• Plan the route of each van in order to keep distances as short as possible • Ensure adequate driving practice and maintenance of vehicles by the contractor

Since most of the personnel have been working at SB for more than ten years and the new operational controls do not present critical difficulties, a single work instruction describing the EMS operational controls was developed and documented. The management team decided to inform all supervisors and workers of these new operational criteria during the upcoming company meetings. During these meetings, top management consulted with the workers and based on their recommendation some signage was added in specific places of the plant to remind the workers of the key elements of these controls.

Knowing that a life cycle perspective was required by ISO 14001, the management team realised the contract established for the new building addressed environmental requirements in the design. They started thinking about the goods and services they procured from suppliers and the use and disposal of their products by their customers. As a result, John began communicating with their main suppliers and customers in order to promote joint actions to protect the environment.

By communicating with customers, Hillary was able to share the potential adverse landfill impacts associated with disposal and make recommendations to recycle the components as scrap metal. By working with their suppliers, John was able to identify some environmental requirements for the key goods and services purchased by SB. As an example, SB had signed a long term contract with the main supplier of packing and packaging materials. Included in the contract conditions was a review of the design requirements with the intention to decrease the weight of these materials without affecting their physical properties, and, if possible, explore the feasibility of materials that can be returned to SB and used more than once.

To comply with local regulations, SB has a fire emergency plan and a spill mitigation plan which it tests regularly. A new emergency plan was developed to respond to channel floods and describe the actions in place to prevent and mitigate damage. These plans not only describe how to respond to an emergency, but also how business may be affected.

11.5 Performance evaluation/compliance/internal audits

Monitoring and measurement of SB's significant environmental aspects, operational controls and environmental objectives and subsequent analysis and evaluation is performed as follows in Table 20.

Table 20 — SB monitoring, measurement, analysis and evaluation schedule

Significant aspect	What (indicators)	Method	Acceptance criteria	When performed	When evaluated
Chemical spills	Availability of response equipment and supplies	Visual inspection	Legal requirements	Monthly	After each inspection
Potential fire					
Energy consumption	% O_2, in exit gaseous emission	Continuous meter, calibrated annually per manufacturer specification	4 % > % O_2 > 8 %	Continuously	Daily checks
	Furnace fuel consumption (Kcal)	Natural gas volume meter calibrated annually per manufacturer specification	10 % reduction in Kcal per tonnes of thermally treated product	Continuous	Monthly
Discharge of water to the channel	Chemical and biological parameters (concentration and number of violations)	Accredited laboratory sampling and analysis methods	Concentration < applicable regulations; zero violations	Every 6 months	Every 6 months
Emission of vapours	Chemical compounds, particulates (concentration and number of violations)	Accredited laboratory sampling and analysis methods	Concentration < applicable regulations; zero violations	Every 6 months	Every 6 months
Hazardous waste generation	Tonnes hazardous waste disposed	Verification of transport and treatment certificates	Certificates that comply with legal requirements	Every batch	Contractor approval; internal audits

Significant aspect	What (indicators)	Method	Acceptance criteria	When performed	When evaluated
Disposal of paper, plastics and wood by the customer	SB does not have full control of this aspect. However, SB tries to influence customers by recommending methods to reuse the materials or, eventually, to return them to SB				
Vans — consumption of fuel	Litres of fuel/100 km travelled	Driver log book readings	Manufacturer specifications	Every time fuel is loaded onto vehicles	Quarterly
Noise	Decibels (db) levels in neighbourhood	Accredited laboratory sampling and analysis methods	< 60 db	Initially weekly; then semi-annually	Post equipment upgrade; management review

All instruments used for these measurements are regularly verified or calibrated as recommended by the equipment manufacturer. In the case of external measurements, SB requires the outsourced laboratory to provide evidence of the calibration status of the equipment used. Since the supervisors are responsible for collection of the data, they find it easy to analyse when it is visually displayed on either control charts, tabulations or trends plots. They will provide these analyses to SB management team members for evaluation at the management review meetings.

To assure SB maintains its compliance status with existing and potentially new regulations, Bill also arranged with experts within the Association to conduct a compliance audit every three years.

Bill attended a course on EMS internal audits run by the Association. John asked Bill to carry out an audit of the full EMS initially, then conduct audits focused on the high risk and problematic areas twice a year. John and Hillary would review the audit reports when issued, and review the effectiveness of the corrective action plans at their management review meetings.

11.6 Competence, awareness and communication

Since most of the workers have been at SB for more than ten years, John considered them "competent" personnel; however, because operations were expanding into new product lines and work practices changing with them, their skills and knowledge needed to be updated. John decided to address environmental training requirements in the Three Year EMS Project Plan.

With regard to the two supervisors and new workers that need to be hired because of the expansion, John decided to seek help from the Association. A human relations consultant assisted John in defining the competence requirements for each position that will be hired.

Every year, John invites all personnel to a barbecue in the local football club facilities. He takes some time to share updates on the business. During the last meeting he presented the following to the staff to be sure they were aware of SB's efforts to protect the environment:

- the new environmental policy;
- SB's overall plan for the implementation of the EMS;
- the eight significant environmental aspects and the related operational controls;
- the compliance obligations;
- their expected contribution to the effectiveness of the EMS; and
- the implication of not conforming with the EMS requirements.

He also encouraged all personnel to recommend ideas for continual improvement.

He is planning to repeat this meeting every year, as a "refresher" for existing personnel and to introduce the EMS to any newly hired personnel.

With reference to internal communication, there is no need to establish any new methods, the existing informal communication channels used in this small company are quite sufficient.

John is responsible for all external communications. He consults with Hillary, Bill and the supervisors before sending out information to double-check that it is correct. He is also responsible for responding to any relevant communication received from external sources, so he keeps records of it in his logbook.

11.7 Documented information

John wanted to have all documented information loaded into the computer network of the plant. This is a simple network that allows the six PCs used by John, Hillary, Bill and the three supervisors to be connected to each other. They collect the following documented information in different logbooks that are filed in a folder called "EMS":

- A logbook under the responsibility of John that contains the following documented information:
  - environmental policy (it is also posted in several places of the plant)
  - scope of the EMS

- environmental objectives
- issues related to the context of SB
- compliance obligations
- risks and opportunities that will be addressed
- external communications
- emergency plans (also posted in relevant locations)
- tests of emergency plans, and any follow-up actions
- results of compliance evaluations
- results of internal audits
- results of management reviews
- renewal of the environmental permit, including the monitoring and measurement results.

- A logbook under the responsibility of each supervisor that contains the following documented information:
  - environmental aspects/impacts under their responsibility
  - significant environmental aspects
  - operational controls for those significant aspects
  - monitoring, measurement and analysis of environmental performance and operational controls
  - evidence of personnel competence under their responsibility
  - the nonconformities identified in their sections
  - corrective action taken.

The supervisor´s logbooks are available to all employees, who have access to the supervisors' PCs using a personal password. John's logbook is available to the three supervisors, Bill and Hillary. John downloads a backup of all four logbooks every week on Friday, before leaving work.

11.8 Leadership

John has been SB´s manager for almost 30 years. Nowadays his leadership style may be perceived as somewhat authoritarian, in spite of all the meetings he has regularly with the three supervisors and with all personnel.

Bill and Hillary will lead the company in a more "modern" style, expecting the supervisors to exert their leadership within their scope of responsibilities.

That is why they are planning to incorporate the three supervisors into an expanded "top management" team.

John defined SB´s environmental policy, and posted it in several places of the plant to communicate it to visitors, contractors and all personnel.

EMS-related responsibilities are clearly defined in the work instructions and known by all employees, as SB is a very small company. To reinforce them, John decided to remind all personnel during the annual meetings. In particular, during the last meeting he clarified the following.

- The importance of fulfilling environmental responsibilities to SB's success, and how those responsibilities have been incorporated in SB's business strategy.
- He takes accountability for the effectiveness of the EMS and will support all personnel in their efforts to contribute to the EMS effectiveness so that it can achieve its intended outcomes.
- He will provide the necessary resources for implementation, maintenance and improvement of the EMS.
- He, Bill and Hillary are responsible for ensuring that the EMS meets ISO 14001 requirements, although each employee must also do their part.
- He welcomes any and all suggestions to further protect the environment or improve the EMS.

11.9 Driving improvement

As mentioned earlier, John, Bill and Hillary currently constitute the "top management" of SB. They meet weekly for lunch in a restaurant nearby to discuss the status of the company, and once per month their discussions focus on various topics related to the EMS. This venue provides a quiet environment that is removed from the challenges and distractions of day-to-day operations so they can conduct the management review of their EMS (Clause 9.3).

John keeps a logbook of the EMS topics and key decisions made, as well as notes on progress made on decisions made at previous meetings. This way he ensures that all topics of management review are treated at least every six months.

To help them keep track of key milestones in the deployment of their EMS Program implementation, they developed a three year project plan during the first management review meeting, as shown in Table 21 below.

Table 21 — SB's Three Year EMS Project Plan

	Year 1	Year 2	Year 3
Analysis of the context	Fully developed	Update when the new plant starts operations	
Compliance obligations	Legal requirements only, including those related to the new plant	Add the voluntary requirements the organization chooses to comply with	Contract with association for triennial compliance audit
Indicators	Mainly reactive indicators of pollutant discharges	Add proactive indicators to measure achievement of intended outcomes	
Training	Bill takes a one year course in environmental technology at the local university. Supervisors take a one day "environmental awareness" course. The three supervisors and all workers will be trained in the new plant technology, including the related environmental aspects.	The new two Supervisors will take a one day "environmental awareness" course. Supervisors and the newly hired workers will be trained in the technology at both plants, including the related environmental aspects.	
Management		John retires and hands over the management to Bill and Hillary. Since two supervisors are also retiring, evaluate if the remaining supervisor is ready to join the "top management" team.	Evaluate if the two new supervisors are ready to be integrated to the "top management" team. This will surely improve their commitment to the EMS.

	Year 1	Year 2	Year 3
Equipment	Identify technological options and funding needs for equipment in new building	Identify technological options and funding needs for existing equipment	
EMS recognition			Apply for third party certification

During the second management review meeting, the initial actions proposed to address the risks and opportunities classified "high" (other than aspects and compliance obligations) were discussed and agreed upon. They realised that that they needed to research some options and gather more information before making a final determination of the actions necessary to address these five issues, so they decided to update these actions during the next (third) management review meeting. See Table 22.

Table 22 — Actions planned to address risk and opportunities

	Action proposed	
External context	Move storage to the ground floor of the new building	Bill will review with the engineering company contracted for the new building how space could be made for the material storage and van parking on the ground floor. Hillary will evaluate the funding needs for this change.
	Design the new building in order to allow space for parking inside SB´s premises	
Internal context	Find successors for the retiring supervisors and ensure a minimum overlap of 6 months	John will assess current workers to identify any that could be upgraded to supervisor. If necessary, Bill will look replacements with an employment agency.
Interested parties	Evaluate if operations can start at 08:00 am	Bill and John will analyse this alternative
	Establish a process to ensure reports are filed on time	Bill will work with the environmental consultant to develop a compliance calendar
	Adhere to the environmental charter of the local Association of Metallurgic Organizations	Bill will analyse the requirement of this charter, identify any additions to ISO 14001 requirements, and propose how to comply with them

Regarding EMS performance evaluation, it was noted that a set of performance indicators are needed. SB decided to review the data the supervisors are generating, and test a few indicators during the first year before finalizing the set of indicators they will use to verify:

- environmental performance has improved (absolute and % change)
- compliance obligations have been met (number of noncompliances)
- environmental objectives have been achieved (% actions complete on time).

During the third monthly meeting, the following two environmental objectives were approved for the first year. See Table 23.

Table 23 — Environmental objectives

Environmental aspect	First year environmental objective
Energy consumption	Reduce energy consumption in the existing furnaces by 10 %
Noise, triggered by the complaint of two neighbours	Reduce noise level in the area surrounding the plants, two closest neighbours down to 60 dB

An action plan was developed and implemented to ensure improvement of two key objectives. The programme established the actions needed to meet the objective, the timetable, the resources needed and the person responsible for each action. As an example, the programme related to the noise reduction in the neighbourhood, is shown below in Table 24.

Table 24 — Plan to achieve environmental objective: reduce noise level in the area surrounding the plants, two closest neighbours down to 60 dB

What will be done	Resources required	Person responsible	When it will be completed
Measure the noise level in at least 4 points twice per week in the next 4 weeks	$$$ to contract the external laboratory	Contract: John Monitor in the field: logistics supervisor	In 30 days
Contract a specialist (via the association) to recommend actions to reduce furnace noise *	$$$ for the consultant	John	In 45 days
Select an option	—	John + Bill	In 60 days
Implement the selected option	$$$ to buy the equipment Install equipment	Bill Production/maintenance supervisor	In 75 days
Re-measure the noise level in at least 4 points twice per week in the next 4 weeks	$$$ to contract the external laboratory	Contract: John Monitor in the field: logistics supervisor	30 days after the equipment is installed
Analyse noise data to determine if further action is needed to achieve the objective: < 60 db	—	John & Bill & specialist	30 days after re-measurement

* Depending on the engineering change recommended to the furnaces (for example, it could be the installation of acoustic fences close to the furnaces blowers or dampers to reduce the vibration of the blowers), associated changes to infrastructure maintenance will be incorporated.

Annex A — Finding more information (references and websites)

International Organization for Standardization

Management standards: https://www.iso.org/management-system-standards.html

ISO/TC 207/SC 1 Webpage: https://committee.iso.org/sites/tc207sc1/home/projects/published/iso-14001---environmental-manage.html

ISO 14001: http://www.iso.org/iso/home/store/catalogue_tc/catalogue_detail.htm?csnumber=60857

ISO Focus+ https://www.iso.org/2015/11/Ref2013.html

ISO brochure on ISO 14001: https://www.iso.org/files/live/sites/isoorg/files/archive/pdf/en/iso_14001_-_key_benefits.pdf

ISO brochure, Introduction to ISO 14001:2015: https://www.iso.org/publication/PUB100371.html

ISO brochure, Environmental management — The ISO 14000 family of International Standards (ISBN 978-92-67-10500-0): https://www.iso.org/publication/PUB100238.html

ISO guide, *The integrated use of management system standards* (ISBN 978-92-67-10473-7): http://www.iso.org/iso/pressrelease.htm?refid=Ref1144

For North American subscribers

US Technical Advisory Group to TC 207, Clarification of the Intent of ISO 14001: http://asq.org/2010/12/iso-14000/clarification-of-intent-of-iso-14001.pdf?SSOForce=Y

Small Business Association: Green Business Practices: https://www.sba.gov/managing-business/running-business/green-business-guide/green-business-practices

US Small Business Environmental Home Page: http://www.smallbiz-enviroweb.org

DOE Industrial Assessment Program: Information on natural resource management for manufacturing http://energy.gov/eere/amo/ta

EnviroReady: For Canadian subscribers allowing self-declaration against an Environmental ready report: http://www.14000registry.com/

For United Kingdom subscribers

NetRegs – environmental guidance for business in UK (Compliance and Environmental Regulations): http://www.netregs.org.uk

For European subscribers

GreenEcoNet: Connecting SME's for a green economy http://www.greeneconet.eu/

Enterprise Europe Network, European
Small Business Portal: http://ec.europa.eu/small-business/most-of-market/environment-business/index_en.htm

For Asian subscribers

Ministry of the Environment (MOE): http://www.env.go.jp/en/index.html

Promoting environment-conscious business activities at MOE: http://www.env.go.jp/en/policy/economy/pecba.html

Energy and Environment Policy at Ministry of Economy, Trade and Industry (METI): http://www.meti.go.jp/english/policy/energy_environment/index.html

For all subscribers

General information offering a global context for environmental management: http://www.unep.org/

International Chamber of Commerce, Business Charter for Sustainable Development: http://www.iisd.org/business/tools/principles_icc.aspx

OECD Sustainable Manufacturing Toolkit: http://www.oecd.org/innovation/green/toolkit/aboutsustainablemanufacturingandthetoolkit.htm

ISO Standards

ISO 14004:2016, *Environmental management systems — General guidelines on implementation*

ISO 14031, *Environmental management — Environmental performance evaluation — Guidelines*

ISO 14006, *Environmental management systems — Guidelines for incorporating ecodesign*

ISO 19011, *Guidelines for auditing management systems*

ISO 31000, *Risk management — Principles and guidelines*

CASCO documents

ISO/IEC 17021-1, *Conformity assessment — Requirements for bodies providing audit and certification of management systems — Part 1: Requirements*

ISO 17021-2, *Conformity assessment — Requirements for bodies providing audit and certification of management systems — Part 2: Competence requirements for auditing and certification of environmental management systems*

ISO 17023:2013, *Conformity assessment — Guidelines for determining the duration of management system certification audits*

ISO/IEC 17050-1, *Conformity assessment — Supplier's declaration of conformity — Part 1: General requirements*

Transition Planning Guidance for ISO 14001:2015: http://www.iaf.nu/upFiles/IAFID10Transition_ISO14001publicationversion27022015.pdf

Other potential sources of information

- Industry associations, chambers of commerce, small business associations
- Environmental organizations
- Government departments
- Accreditation/certification/registration bodies
- National standards bodies

Awareness of environmental problems is growing and it is believed that current development patterns cannot be sustained in the long term. The benefits of addressing environmental issues are not only linked to the protection of the environment, but also to business performance and profitability.

This handbook, devised specifically for SMEs, has been updated to take into account the latest edition of the standard published in 2015. It aims to help small businesses understand the requirements of an environmental management system and to help them implement ISO 14001 successfully.

International Organization for Standardization

ISO Central Secretariat
Ch. de Blandonnet 8
CP 401
CH – 1214 Vernier, Geneva
Switzerland

iso.org

We care about our planet.
This handbook is printed on recycled paper
If your library is too full to keep it forever, please recycle it too!

ISBN 978-92-67-10679-3

Made in United States
Orlando, FL
06 November 2026

90240677R00133